Accurate
Condensed-Phase
Quantum Chemistry

COMPUTATION IN CHEMISTRY

Series Editor

Professor Peter Gill
Research School of Chemistry
Australian National University
pgill@rsc.anu.edu.au

Accurate Condensed-Phase Quantum Chemistry

Edited by
Frederick R. Manby

CRC Press
Taylor & Francis Group
Boca Raton London New York

CRC Press is an imprint of the
Taylor & Francis Group, an **informa** business

CRC Press
Taylor & Francis Group
6000 Broken Sound Parkway NW, Suite 300
Boca Raton, FL 33487-2742

First issued in paperback 2019

© 2011 by Taylor and Francis Group, LLC
CRC Press is an imprint of Taylor & Francis Group, an Informa business

No claim to original U.S. Government works

ISBN-13: 978-1-4398-0836-8 (hbk)
ISBN-13: 978-1-138-37414-0 (pbk)

Library of Congress Cataloging-in-Publication Data

Accurate condensed-phase quantum chemistry / editor, Frederick R. Manby.
    p. cm. -- (Computation in chemistry)
  Includes bibliographical references and index.
  ISBN 978-1-4398-0836-8 (hardcover : alk. paper)
  1. Quantum chemistry. 2. Condensed matter. I. Manby, Frederick R.

QD462.A33 2011
541'.28--dc22   2010022634

Visit the Taylor & Francis Web site at
http://www.taylorandfrancis.com

and the CRC Press Web site at
http://www.crcpress.com

Contents

Series Preface

Computational chemistry is highly interdisciplinary, nestling in the fertile region where chemistry meets mathematics, physics, biology, and computer science. Its goal is the prediction of chemical structures, bondings, reactivities, and properties through calculations *in silico*, rather than experiments *in vitro* or *in vivo*. In recent years, it has established a secure place in the undergraduate curriculum and modern graduates are increasingly familiar with the theory and practice of this subject. In the twenty-first century, as the prices of chemicals increase, governments enact ever-stricter safety legislations, and the performance/price ratios of computers increase, it is certain that computational chemistry will become an increasingly attractive and viable partner of experiment.

However, the relatively recent and sudden arrival of this subject has not been unproblematic. As the technical vocabulary of computational chemistry has grown and evolved, a serious language barrier has developed between those who prepare new methods and those who use them to tackle real chemical problems. There are only a few good textbooks; the subject continues to advance at a prodigious pace and it is clear that the daily practice of the community as a whole lags many years behind the state of the art. The field continues to advance and many topics that require detailed development are unsuitable for publication in a journal because of space limitations. Recent advances are available within complicated software programs but the average practitioner struggles to find helpful guidance through the growing maze of such packages.

This has prompted us to develop a series of books entitled *Computation in Chemistry* that aims to address these pressing issues, presenting specific topics in computational chemistry for a wide audience. The scope of this series is broad, and encompasses all the important topics that constitute "computational chemistry" as generally understood by chemists. The books' authors are leading scientists from around the world, chosen on the basis of their acknowledged expertise and their communication skills. Where topics overlap with fraternal disciplines—for example, quantum mechanics (physics) or computer-based drug design (pharmacology)—the treatment aims primarily to be accessible to, and serve the needs of, chemists.

This book, the second in the series, brings together recent advances in the accurate quantum mechanical treatment of condensed systems, whether periodic or aperiodic, solid or liquid. The authors, who include leading figures from both sides of the Atlantic, describe methods by which the established methods of gas-phase quantum chemistry can be modified and generalized into forms suitable for application to extended systems and, in doing so, open a range of exciting new possibilities for the subject. Each chapter exemplifies the overarching principle of this series: These will not be dusty technical monographs but, rather, books that will sit on every practitioner's desk.

Preface

Quantum mechanical calculations on polyatomic molecules are necessarily approximate. But through the development of hierarchies of approximate treatments of the electron correlation problem, accuracy can be systematically improved. This book explores several attempts to apply the successful methods of molecular electronic structure theory to condensed-phase systems, and in particular to molecular liquids and crystalline solids.

The wavefunction-based methods described in this book all begin with a mean-field calculation to produce the Hartree–Fock energy. Relativistic effects are neglected and the Born–Oppenheimer approximation is assumed. The remaining part of the electronic energy arises from electron correlation. Neither of these terms is easy to compute in periodic boundary conditions.

The Hartree–Fock theory for crystalline solids has a distinguished history, starting with expansion of the molecular orbitals as a linear combination of (Gaussian-type) atomic orbitals [1], leading, for example, to the development of the CRYSTAL code [2]. CRYSTAL is perhaps the most extensively tested implementation of periodic Hartree–Fock theory, and the accuracy of the whole approach has recently been greatly extended by the development of periodic second-order Møller–Plesset perturbation theory (MP2) in the CRYSCOR collaboration (see [3] and Chapter 2). This allows for accurate, correlated treatments of complex materials, and high computational efficiency is achieved through a combination of density fitting and local treatments of electron correlation.

Periodic Hartree–Fock using Gaussian-type orbitals has also been implemented by the Scuseria group in the GAUSSIAN electronic structure package [4] (a recent paper on the periodic Hartree–Fock implementation can be found in [5]), and they too have developed periodic MP2 methods, based on an atomic-orbital-driven Laplace-transform formalism of the theory (see Chapter 1 and references therein). This has recently been further accelerated through the introduction of density fitting (or resolution of the identity) techniques, as described in Chapter 1.

The alternative approach for periodic Hartree–Fock theory is to represent the molecular orbitals in a basis set of plane waves. This, in combination with pseudopotentials for the effective description of core electrons, has proven extraordinarily successful for periodic density functional theory (DFT), because the Coulomb energy, which is a major challenge for atomic-orbital methods, can be evaluated extremely easily. There are implementations of various flavors of the approach in various codes, including VASP [6], CASTEP [7], PWSCF [8] and CP2K [9].

As part of an attempt in Bristol and University College London to characterize a simple crystalline solid—lithium hydride—as accurately as possible [10, 11], Gillan et al. invested very considerable effort in determining accurate Hartree–Fock energies for the crystal [12]. This proved extraordinarily difficult, and it was found that converging atomic-orbital-based Hartree–Fock calculations to high accuracy was extremely difficult, because linear dependence problems made it impossible to make the basis set sufficiently flexible.* Instead, the approach used was based on plane-wave Hartree–Fock calculations with corrections to remove the effect of the pseudopotentials. This yielded a static cohesive energy at $a = 4.084$ Å of -131.95 mE_h [12]. This test system has subsequently been studied in three groups, producing Hartree–Fock results with an amazing degree of agreement (see Section 4.3).

The wavefunction-based treatment of electron correlation for crystalline solids is also an area of very considerable activity. Various approaches have been devised that attack the problem directly: examples include the periodic atomic-orbital-based MP2 method developed in the Scuseria group (see Chapter 1); the local and density-fitted MP2 approach of the Regensburg and Torino groups (see Chapter 2); and the plane-wave-based MP2 code developed in Vasp [13]. In fact, in Vasp, methods beyond MP2 are being developed including the random-phase approximation [14] and even coupled-cluster theory. Hirata describes work in his group on periodic electronic structure theory using localized and crystal orbital approaches in Chapter 6.

An alternative approach is to treat electron correlation through considering finite clusters. This is the principle at the heart of both the hierarchical scheme (Chapter 4) and of the various approaches inspired by the many-body expansion (Chapters 3 and 5–7). In the hierarchical scheme the surface effects that dominate the properties of small clusters are carefully removed to reveal properties that accurately reflect the bulk solid (Chapter 4). In the incremental scheme, one of the oldest and best tested approaches for the wavefunction-based treatment of electron correlation in solids, a periodic Hartree–Fock calculation is followed by a many-body expansion of the correlation energy, where the individual units of the expansion are either atoms or other domains of localized molecular orbitals (Chapter 3).

In the work of Dahlke Speetzen et al. (Chapter 5), the many-body expansion of the energy of a molecular cluster is made more rapidly convergent through embedding lower-order contributions in suitable point charge representations of the remaining molecules. These authors also explore the feasibility of applying their methodology to Monte Carlo

---

* It is only fair to note that this problem can be avoided and highly converged orbital-based Hartree–Fock theory is certainly possible—see, for example, [5].

simulations of bulk molecular liquids. Hirata describes another many-body approach, aimed at periodic systems, in which the fragments are embedded in an electrostatic representation of the remainder of the system, which is self-consistently optimized (Chapter 6). Through development of analytic derivatives for this scheme, Hirata has been able to compute optimized structures, phonon dispersion curves, and Raman spectra for extended systems, and an overview of this work is presented here. Finally, O'Neill et al. describe a many-body expansion technique aimed directly at the simulation of molecular liquids, and present MP2-level radial distribution functions for liquid water (Chapter 7).

Overall, a trend is emerging: where previously Hartree–Fock and DFT calculations (and perhaps quantum Monte Carlo) were the only feasible options for treating the electronic structure of condensed-phase systems, it is now possible to treat crystals with MP2 and coupled-cluster theory, and it is becoming possible to simulate liquids using wavefunction-based electronic structure theory. The chapters gathered in this volume cover a wide range of exciting and novel approaches for theoretical treatment of solids and liquids, and constitute some of the first steps toward accurate, and systematically improvable, quantum chemistry for condensed phases.

Frederick R. Manby
Centre for Computational Chemistry, School of Chemistry
University of Bristol, Bristol, U.K.

References

[1] M. Causà, R. Dovesi, C. Pisani, and C. Roetti. Electronic structure and stability of different crystal phases of magnesium oxide. *Phys. Rev.* B 33, 1308 (1986).

[2] C. Pisani, R. Dovesi, C. Roetti, M. Causà, R. Orlando, S. Casassa, and V. R. Saunders. CRYSTAL and EMBED, two computational tools for the ab initio study of electronic properties of crystals. *Int. J. Quantum Chem.* 77, 1032 (2000).

[3] L. Maschio, D. Usvyat, F. R. Manby, S. Cassassa, C. Pisani, and M. Schütz. Fast local-MP2 method with density-fitting for crystals. I. Theory. *Phys. Rev.* B 76, 075101 (2007).

[4] M. J. Frisch, G. W. Trucks, H. B. Schlegel, G. E. Scuseria, M. A. Robb, J. R. Cheeseman, G. Scalmani, V. Barone, B. Mennucci, G. A. Petersson, H. Nakatsuji, M. Caricato, X. Li, H. P. Hratchian, A. F. Izmaylov, J. Bloino, G. Zheng, J. L. Sonnenberg, M. Hada, M. Ehara, K. Toyota, R. Fukuda, J. Hasegawa, M. Ishida, T. Nakajima, Y. Honda, O. Kitao, H. Nakai, T. Vreven, Montgomery, Jr., J. A., J. E. Peralta, F. Ogliaro, M. Bearpark, J. J. Heyd, E. Brothers, K. N. Kudin, V. N. Staroverov, R. Kobayashi, J. Normand, K. Raghavachari, A. Rendell, J. C. Burant, S. S. Iyengar, J. Tomasi, M. Cossi, N. Rega, J. M. Millam, M. Klene, J. E. Knox, J. B. Cross, V. Bakken, C. Adamo, J. Jaramillo, R. Gomperts, R. E. Stratmann, O. Yazyev, A. J. Austin, R. Cammi,

C. Pomelli, J. W. Ochterski, R. L. Martin, K. Morokuma, V. G. Zakrzewski, G. A. Voth, P. Salvador, J. J. Dannenberg, S. Dapprich, A. D. Daniels, O. Farkas, J. B. Foresman, J. V. Ortiz, J. Cioslowski, and D. J. Fox. GAUSSIAN 09 Revision A.1, (Gaussian Inc., Wallingford CT, 2009).

[5] J. Paier, C. V. Diaconu, G. E. Scuseria, M. Guidon, J. VandeVondele, and J. Hutter. Accurate Hartree–Fock energy of extended systems using large Gaussian basis sets. *Phys. Rev.* B 80, 174114 (2009).

[6] J. Paier, R. Hirschl, M. Marsman, and G. Kresse. The Perdew–Burke–Ernzerhof exchange-correlation functional applied to the G2-1 test set using a plane-wave basis set. *J. Chem. Phys.* 122, 234102 (2005).

[7] S. J. Clark, M. D. Segall, C. J. Pickard, P. J. Hasnip, M. I. J. Probert, K. Refson, and M. C. Payne. First principles methods using CASTEP. *Z. Kristallogr.* 220, 567 (2005).

[8] S. Scandolo, P. Giannozzi, C. Cavazzoni, S. de Gironcoli, A. Pasquarello, and S. Baroni. First-principles codes for computational crystallography in the Quantum-ESPRESSO package. *Z. Kristallogr.* 220, 574 (2005).

[9] M. Guidon, J. Hutter, and J. VandeVondele. Robust periodic Hartree–Fock exchange for large-scale simulations using Gaussian basis sets. *J. Chem. Theo. Comp.* 5, 3010 (2009).

[10] F. R. Manby, D. Alfè, and M. J. Gillan. Extension of molecular electronic structure methods to the solid state: computation of the cohesive energy of lithium hydride. *Phys. Chem. Chem. Phys.* 8, 5178 (2006).

[11] S. J. Nolan, M. J. Gillan, D. Alfè, N. L. Allan, and F. R. Manby. Comparison of the incremental and hierarchical methods for crystalline neon. *Phys. Rev.* B 80, 165109 (2009).

[12] M. J. Gillan, F. R. Manby, D. Alfè, and S. de Gironcoli. High-precision calculation of Hartree–Fock energy of crystals. *J. Comput. Chem.* 29, 2098 (2008).

[13] M. Marsman, A. Grüneis, J. Paier, and G. Kresse. Second-order Møller–Plesset perturbation theory applied to extended systems. I. Within the projector-augmented-wave formalism using a plane wave basis set. *J. Chem. Phys.* 130, 184103 (2009).

[14] J. Harl and G. Kresse. Accurate bulk properties from approximate many-body techniques. *Phys. Rev. Lett.* 103, 056401 (2009).

Editor

Frederick R. Manby is a Reader in the Centre for Computational Chemistry in the School of Chemistry at the University of Bristol, and was previously a Royal Society University Research Fellow. His research has focused on two main areas: first, on development of efficient and accurate electronic structure methods for large molecules. Second, he has worked on accurate treatment of condensed-phase systems, including electron correlation in crystalline solids, and on application of wavefunction-based electronic structure theories for molecular liquids, particularly water. He has been awarded the Annual Medal of the International Academy of Quantum Molecular Sciences (2007) and the Marlow Medal of the Royal Society of Chemistry (2006) for his research in molecular electronic structure theory.

Contributors

Neil L. Allan
Centre for Computational
  Chemistry
School of Chemistry
University of Bristol
Bristol, U.K.

Simon Binnie
London Centre for
  Nanotechnology
Department of Physics
  and Astronomy
University College London
London, U.K.

Peter Bygrave
Centre for Computational
  Chemistry
School of Chemistry
University of Bristol
Bristol, U.K.

Silvia Casassa
Dipartimento di Chimica
  and Centre of Excellence
Nanostructured Interfaces
  and Surfaces
Università di Torino
Torino, Italy

Michael J. Gillan
Department of Physics
  and Astronomy
University College London
London, U.K.

Migen Halo
Dipartimento di Chimica
  and Centre of Excellence
Nanostructured Interfaces
  and Surfaces
Università di Torino
Torino, Italy

So Hirata
Quantum Theory Project
University of Florida
Gainesville, Florida

Artur F. Izmaylov
Department of Chemistry
Yale University
New Haven, Connecticut

Murat Keçeli
Quantum Theory Project
  and Center for Macromolecular
  Science and Engineering
Department of Chemistry and
  Department of Physics
University of Florida
Gainesville, Florida

Hannah R. Leverentz
Department of Chemistry
  and Supercomputing
  Institute
University of Minnesota
Minneapolis, Minnesota

Hai Lin
Chemistry Department
University of Colorado Denver
Denver, Colorado

Marco Lorenz
Institute for Physical and
  Theoretical Chemistry
Universität Regensburg
Regensburg, Germany

Frederick R. Manby
Centre for Computational
  Chemistry
School of Chemistry
University of Bristol
Bristol, U.K.

Lorenzo Maschio
Dipartimento di Chimica
  and Centre of Excellence
Nanostructured Interfaces
  and Surfaces
Università di Torino
Torino, Italy

Stephen Nolan
Centre for Computational
  Chemistry
School of Chemistry
University of Bristol
Bristol, U.K.

Darragh P. O'Neill
Centre for Computational
  Chemistry
School of Chemistry
University of Bristol
Bristol, U.K.

Beate Paulus
Institut für Chemie und Biochemie
Freie Universität Berlin
Berlin, Germany

Cesare Pisani
Dipartimento di Chimica
  and Centre of Excellence
Nanostructured Interfaces
  and Surfaces
Università di Torino
Torino, Italy

Martin Schütz
Institute for Physical and
  Theoretical Chemistry
Universität Regensburg
Regensburg, Germany

Gustavo E. Scuseria
Department of Chemistry
Rice University
Houston, Texas

Tomomi Shimazaki
Fracture and Reliability Research
  Institute
Graduate School of Engineering
Tohoku University
Sendai, Japan

Olaseni Sode
Quantum Theory Project
University of Florida
Gainesville, Florida

Erin Dahlke Speetzen
Chemistry Program, Division
  of Life and Molecular Sciences
Loras College
Dubuque, Iowa

Hermann Stoll
Institut für Theoretische Chemie
Universität Stuttgart
Stuttgart, Germany

Donald G. Truhlar
Department of Chemistry
  and Supercomputing Institute
University of Minnesota
Minneapolis, Minnesota

Denis Usvyat
Institute for Physical and
  Theoretical Chemistry
Universität Regensburg
Regensburg, Germany

chapter one

Laplace transform second-order Møller–Plesset methods in the atomic orbital basis for periodic systems

Artur F. Izmaylov and Gustavo E. Scuseria

Contents

1.1 Introduction

The electron correlation energy is much smaller than the Hartree–Fock (HF) energy. However, it is of crucial importance for modeling the electronic structure and properties of molecules and solids. The most popular approaches for including electron correlation are density functional theory (DFT) and wavefunction methods. DFT usually yields a very good value in terms of accuracy over computational cost. Unfortunately, there is no

straightforward path in DFT to get "the right answer for the right reason."
The latter should be interpreted as a series of well-controlled approximations leading to the exact answer. Traditional semilocal DFT also has problems accurately describing dispersion interactions and transition states.
On the other hand, wavefunction methods do yield a straightforward and
systematic way of improving accuracy, although their computational cost
is usually much higher than that of DFT.

The simplest wavefunction approach to the electron correlation problem is second-order Møller–Plesset perturbation theory (MP2). MP2 radi-
cally improves upon HF for dispersion interactions [1], barrier heights [2],
and nuclear magnetic resonance shifts [3] in molecules, and band gaps and
equilibrium geometries in periodic systems [4,5].

The formal scaling of MP2 in traditional, delocalized, canonical orbital
bases is $O(N^5)$, where N is a parameter proportional to the system size [6].
This steep computational cost can be drastically reduced by using local
MP2 method formulations [7–10] (see also Chapter 2), or the atomic orbital Laplace transformed MP2 method (AO-LT-MP2) [11, 12]. While the
former approach uses localized orbitals, the latter exploits the natural locality of atomic orbitals. Both of these formulations provide asymptotic
$O(N)$ computational scaling. The latter has been generalized for periodic
systems [13,14]. Computational cost is not only determined by scaling but
also by prefactors. The main computational bottleneck in MP2 linear scaling methods is the transformation of two-electron integrals. Two-electron
integrals are essentially four-center terms whose evaluation and transformation have $O(N^4)$ and $O(N^5)$ complexity, respectively.

Integral generation can be reduced to quadratic scaling using Cauchy–
Schwarz screening [15], and even to linear scaling if multipole-moment-
based thresholding is applied [12, 16–18]. The application of screening
protocols in the local MP2 and AO-LT-MP2 methods reduces the integral transformation step from $O(N^5)$ complexity to $O(N)$. However, the
prefactor is still large, and to efficiently exploit the local nature of correlation (*nearsightedness principle*), the system under consideration must
be fairly large. In order to obtain an even smaller prefactor, resolution of
the identity (RI) or density-fitting procedures may be introduced. These
are robust alternatives substituting a pair of basis functions in the bra
or ket part of a two-electron integral by a single fitting function [19].
Application of the RI technique to the MP2 formulation leads to an energy expression with three-center, two-electron integrals. This reduces
the complexity of the integral generation and transformation by one
order of magnitude. Although the RI procedure itself has $O(N^3)$ scaling, its prefactor is very small. The RI expansion has been introduced
into local MP2 and AO-LT-MP2 procedures for systems with periodic
boundary conditions (PBC). These techniques significantly accelerate the

computational speed with only a minor loss of accuracy [20–22]; see also Chapter 2.

In this chapter we review the AO-LT-MP2 and RI-AO-LT-MP2 methods with a special emphasis on algorithmic features that are responsible for the computational efficiency of these methods. A comparative assessment of two methods and some illustrative examples of band gap calculations will be given at the end.

1.2 Method

Our consideration will be restricted to the closed-shell case because its generalization to the open-shell case is quite straightforward. In this chapter we use Gaussian atomic orbitals (AOs)

$$\mu_p(\mathbf{r}) = (x - R_x)^l (y - R_y)^m (z - R_z)^n \times \exp\left[-\eta(\mathbf{r} - \mathbf{R} - \mathbf{p})^2\right], \qquad (1.1)$$

where (l, m, n) are integers determining the orbital angular momentum, η is the orbital exponent, and $\mathbf{R} = (R_x, R_y, R_z)$ are the coordinates of the AO center in the unit cell $\mathbf{p}$. In periodic case, HF self-consistent field (SCF) crystal orbitals (CO) are linear combinations of AOs that satisfy the Bloch theorem [23]

$$j_k(\mathbf{r}) = N_c^{-1/2} \sum_{u=0}^{N_c} \sum_{\mu=1}^{N_0} \mu_u(\mathbf{r}) C(\mathbf{k})_{\mu j} e^{i u k}, \qquad (1.2)$$

where $C(\mathbf{k})_{\mu j}$ are CO coefficients, N_0 is the number of AOs per unit cell, and N_c is the number of unit cells. Throughout this chapter we use Greek letters for AOs, Roman letters $i, j, \ldots$ for occupied and $a, b, \ldots$ for virtual COs, $K, L, \ldots$ and $p, q, \ldots$ for the RI basis set and translational vectors. Using HF COs and the Mulliken integral notation

$$(i_{k_1} a_{k_3} | j_{k_2} b_{k_4}) = \int d\mathbf{r}_1 \int d\mathbf{r}_2 \frac{i_{k_1}^*(\mathbf{r}_1) a_{k_3}(\mathbf{r}_1) j_{k_2}^*(\mathbf{r}_2) b_{k_4}(\mathbf{r}_2)}{|r_1 - r_2|}, \qquad (1.3)$$

the MP2 correlation energy per unit cell is

$$E^{\mathrm{MP2}}$$

$$= \mathrm{Re}\left[\frac{1}{V_k^4} \int d\mathbf{k}_{1-4} \sum_{ij,ab} \frac{(i_{k_1} a_{k_3} | j_{k_2} b_{k_4})[2(i_{k_1} a_{k_3} | j_{k_2} b_{k_4}) - (i_{k_1} b_{k_4} | j_{k_2} a_{k_3})]^*}{\epsilon_i(\mathbf{k}_1) + \epsilon_j(\mathbf{k}_2) - \epsilon_a(\mathbf{k}_3) - \epsilon_b(\mathbf{k}_4)}\right], \qquad (1.4)$$

where $\epsilon(\mathbf{k})$ is an HF orbital energy, and V_k is the volume of the Brillouin zone. The MP2 band gap expression can be derived by considering the second-order self-energy correction to the gth HF orbital energy [4, 24, 25]

$$\epsilon_g^{\text{MP2}}(\mathbf{k}) = \epsilon_g^{\text{HF}}(\mathbf{k}) + U(g, \mathbf{k}) + V(g, \mathbf{k}), \tag{1.5}$$

where

$$U(g, \mathbf{k}) = \text{Re}\left[\frac{1}{V_k^3} \int d\mathbf{k}_{1-3} \sum_{i,ab} \frac{(i_{k_1} a_{k_3} | g_k b_{k_2})[2(i_{k_1} a_{k_3} | g_k b_{k_2}) - (i_{k_1} b_{k_2} | g_k a_{k_3})]^*}{\epsilon_i(\mathbf{k}_1) + \epsilon_g(\mathbf{k}) - \epsilon_a(\mathbf{k}_3) - \epsilon_b(\mathbf{k}_2)}\right], \tag{1.6}$$

$$V(g, \mathbf{k}) = \text{Re}\left[\frac{1}{V_k^3} \int d\mathbf{k}_{1-3} \sum_{i,ab} \frac{(i_{k_1} a_{k_3} | j_{k_2} g_k)[2(i_{k_1} a_{k_3} | j_{k_2} g_k) - (i_{k_1} g_k | j_{k_2} a_{k_3})]^*}{\epsilon_i(\mathbf{k}_1) + \epsilon_j(\mathbf{k}_2) - \epsilon_a(\mathbf{k}_3) - \epsilon_g(\mathbf{k})}\right]. \tag{1.7}$$

Then the MP2 correction to the HF direct fundamental band gap is the difference between self-energy corrections to the highest occupied CO (HOCO) and the lowest unoccupied CO (LUCO)

$$\begin{aligned} E_g^{\text{MP2}} = &\, U(\text{HOCO}, \mathbf{k}_{\min}) - U(\text{LUCO}, \mathbf{k}_{\min}) \\ &+ V(\text{HOCO}, \mathbf{k}_{\min}) - V(\text{LUCO}, \mathbf{k}_{\min}), \end{aligned} \tag{1.8}$$

where k-point $\mathbf{k}_{\min}$ minimizes the total band gap. The fundamental gap is an energy difference between the electron attachment and detachment processes, and it must not be confused with the optical gap, which represents the lowest electronic excitation [26].

In addition to computational difficulties related to the delocalized character of canonical COs in Equations (1.4), (1.6), (1.7), an entanglement of different $\mathbf{k}$ vectors in orbital energy denominators requires a computationally expensive multidimensional k-integration. A simple and elegant way to decouple different $\mathbf{k}$ vectors is to apply the Laplace transform to the energy denominators [11, 27]

$$\frac{1}{\epsilon_i(\mathbf{k}_1) + \epsilon_j(\mathbf{k}_2) - \epsilon_a(\mathbf{k}_3) - \epsilon_b(\mathbf{k}_4)} = -\int_0^\infty dt\, e^{[\epsilon_i(\mathbf{k}_1) + \epsilon_j(\mathbf{k}_2)]t} e^{-[\epsilon_a(\mathbf{k}_3) + \epsilon_b(\mathbf{k}_4)]t}. \tag{1.9}$$

This identity is valid only when the denominator preserves its sign for all $\mathbf{k}$ vectors. Thus, it can be used within the MP2 method which is applicable only to the systems where $\epsilon_i(\mathbf{k}_1) + \epsilon_j(\mathbf{k}_2) < \epsilon_a(\mathbf{k}_3) + \epsilon_b(\mathbf{k}_4)$. After an appropriate discretization of the Laplace integral [28]

$$\int_0^\infty dt\, e^{[\epsilon_i(\mathbf{k}_1) + \epsilon_j(\mathbf{k}_2)]t} e^{-[\epsilon_a(\mathbf{k}_3) + \epsilon_b(\mathbf{k}_4)]t} \rightarrow \sum_{t=1}^{N_t} w_t e^{[\epsilon_i(\mathbf{k}_1) + \epsilon_j(\mathbf{k}_2)]t} e^{-[\epsilon_a(\mathbf{k}_3) + \epsilon_b(\mathbf{k}_4)]t}, \tag{1.10}$$

we can rewrite the MP2 energy and band gap correction (Equations [1.4] and [1.8]) in the AO form

$$E^{\text{MP2}} = \sum_{t=1}^{N_t} \sum_{\mu\nu\lambda\sigma,prs} T_{\mu_0\lambda_r}^{\nu_p\sigma_s}(t)[2(\mu_0\nu_p|\lambda_r\sigma_s) - (\mu_0\sigma_s|\lambda_r\nu_p)]$$

$$= \sum_{t=1}^{N_t} \sum_{\mu\nu\lambda\sigma,prs} (\mu_0\nu_p|\lambda_r\sigma_s)\big[2T_{\mu_0\lambda_r}^{\nu_p\sigma_s}(t) - T_{\mu_0\lambda_r}^{\sigma_s\nu_p}(t)\big], \tag{1.11}$$

and

$$E_g^{\text{MP2}} = \sum_{t=1}^{N_t} \sum_{\mu\nu\lambda\sigma,prs} G_{\mu_0\lambda_r}^{\nu_p\sigma_s}(t)[2(\mu_0\nu_p|\lambda_r\sigma_s) - (\mu_0\sigma_s|\lambda_r\nu_p)]$$

$$= \sum_{t=1}^{N_t} \sum_{\mu\nu\lambda\sigma,prs} (\mu_0\nu_p|\lambda_r\sigma_s)\big[2G_{\mu_0\lambda_r}^{\nu_p\sigma_s}(t) - G_{\mu_0\lambda_r}^{\sigma_s\nu_p}(t)\big]. \tag{1.12}$$

The tensors T and G in Equations (1.11) and (1.12) are transformed Coulomb two-electron integrals

$$T_{\mu_0\lambda_r}^{\nu_p\sigma_s}(t) = \sum_{\gamma\delta\kappa\tau,quvw} X_{\mu_0\gamma_q}^t Y_{\nu_p\delta_u}^t (\gamma_q\delta_u|\kappa_v\tau_w) X_{\lambda_r\kappa_v}^t Y_{\sigma_s\tau_w}^t$$

$$= (\mu_0\overline{\nu_p}|\lambda_q\overline{\sigma_s}), \tag{1.13}$$

$$G_{\mu_0\lambda_r}^{\nu_p\sigma_s}(t) = \sum_{\gamma\delta\kappa\tau,quvw} \big[W_{\mu_0\gamma_q}^t Y_{\nu_p\delta_u}^t + X_{\mu_0\gamma_q}^t Z_{\nu_p\delta_u}^t\big](\gamma_q\delta_u|\kappa_v\tau_w) X_{\lambda_r\kappa_v}^t Y_{\sigma_s\tau_w}^t, \tag{1.14}$$

where matrices $\mathbf{X}$, $\mathbf{Y}$, $\mathbf{W}$, and $\mathbf{Z}$ have a form of weighted densities

$$X_{\mu_p\gamma_s}^t = \frac{w_t^{1/4}}{V_k} \int d\mathbf{k} \sum_j C(\mathbf{k})_{\mu j}^* e^{-\epsilon_j(\mathbf{k})t} C(\mathbf{k})_{\gamma j} e^{i\mathbf{k}(\mathbf{p}-\mathbf{s})}, \tag{1.15}$$

$$Y_{\mu_p\gamma_s}^t = \frac{w_t^{1/4}}{V_k} \int d\mathbf{k} \sum_a C(\mathbf{k})_{\mu a}^* e^{\epsilon_a(\mathbf{k})t} C(\mathbf{k})_{\gamma a} e^{i\mathbf{k}(\mathbf{p}-\mathbf{s})}, \tag{1.16}$$

$$W_{\mu_p\gamma_s}^t = w_t^{1/4}\big[C(\mathbf{k})_{\mu g}^* e^{-\epsilon_g(\mathbf{k})t} C(\mathbf{k})_{\gamma g} - C(\mathbf{k})_{\mu g'}^* e^{-\epsilon_{g'}(\mathbf{k})t} C(\mathbf{k})_{\gamma g'}\big] e^{i\mathbf{k}(\mathbf{p}-\mathbf{s})}, \tag{1.17}$$

$$Z_{\mu_p\gamma_s}^t = w_t^{1/4}\big[C(\mathbf{k})_{\mu g'}^* e^{\epsilon_{g'}(\mathbf{k})t} C(\mathbf{k})_{\gamma g'} - C(\mathbf{k})_{\mu g}^* e^{\epsilon_g(\mathbf{k})t} C(\mathbf{k})_{\gamma g}\big] e^{i\mathbf{k}(\mathbf{p}-\mathbf{s})}, \tag{1.18}$$

with g and g', respectively, HOCO and LUCO. Comparison of Equations (1.11) and (1.12) for the MP2 energy and band gap correction to the canonical reciprocal-space Equations (1.4), (1.6), and (1.7) reveals that the multidimensional k-integration has been reduced to a series of independent Fourier transforms (Equations [1.15]–[1.18]). The evaluation of the $\mathbf{X}$,

Y, **W**, and **Z** matrices is the only part of the AO-LT-MP2 calculation that does depend on the number of k-points (N_k) employed in the discretization of the Brillouin zone [29]. Therefore, the computational cost of the AO-LT-MP2 method is essentially N_k-independent, because the CPU time for the **X**, **Y**, **W**, and **Z** construction is negligible with respect to the total CPU time of the AO-LT-MP2 calculation.

```
Generation: (μ₀νₚ|λᵣσₛ) [O(N³_c N⁴_0)]
Generation: (μ₀νₚ||λᵣσₛ) = 2(μ₀νₚ|λᵣσₛ) − (μ₀σₛ|λᵣνₚ) [O(N³_c N⁴_0)]
Loop over Laplace points t = 1, N_t
    Generation: (γ_q δ_u|κ_v τ_w) [O(N⁴)]
    1st transformation: (μ₀δ_u|κ_v τ_w) = Σ X^t_{μ₀γ_q}(γ_q δ_u|κ_v τ_w)[O(N₀N⁴)]
    2nd transformation: (μ₀ν̄_q|κ_v τ_w) = Σ Y^t_{ν_q δ_u}(μ₀δ_u|κ_v τ_w)[O(N₀N⁴)]
    3rd transformation: (μ₀ν̄_q|λᵣτ_w) = Σ X^t_{λᵣκ_v}(μ₀ν̄_q|κ_v τ_w)[O(N₀N⁴)]
    4th transformation: (μ₀ν̄_q|λᵣσ̄_s) = Σ Y^t_{σ_s τ_w}(μ₀ν̄_q|λᵣτ_w)[O(N₀N⁴)]
    Contraction: e_t = Σ(μ₀ν̄_p|λ_q σ̄_s)(μ₀νₚ||λ_q σₛ) [O(N₀N³)]
End loop over t
```

Scheme 1.1

The general flow of the AO-LT-MP2 algorithm is presented in Scheme 1.1, where the formal complexity of each step is presented in square brackets, and N is the product $N_c N_0$. The limiting steps of the AO-LT-MP2 formulation are integral transformations (Equations [1.13] and [1.14]); their formal scaling is $O(N_0 N^4)$. In order to improve upon the formal scaling of each step, various integral screening protocols were added to Scheme 1.1 [13]. Besides screening of two-electron integrals, which will be considered in detail later, we also can employ the RI approximation to two-electron integrals

$$(\mu_0 \nu_p | \lambda_q \sigma_r) \approx \sum_{\overline{K_s}, \overline{L_u}} (\mu_0 \nu_p | \overline{K_s}) A^{-1}_{\overline{K_s L_u}} (\overline{L_u} | \lambda_q \sigma_r), \tag{1.19}$$

$$A_{\overline{K_s L_u}} = \int d\mathbf{r_1} \int d\mathbf{r_2} \frac{\overline{K_s(\mathbf{r_1})} \overline{L_u(\mathbf{r_2})}}{|r_1 - r_2|}. \tag{1.20}$$

To obtain a symmetric representation, the matrix **A** is decomposed and its parts are used for a transition to the orthonormal RI basis $\{K, L, \dots\}$

$$\sum_{K,L,su} (\mu_0 \nu_p | \overline{K_s}) A^{-1}_{\overline{K_s L_u}} (\overline{L_u} | \lambda_q \sigma_r) = \sum_{K,L,M,suv} (\mu_0 \nu_p | \overline{K}) A^{-1/2}_{\overline{K_s M_u}} A^{-1/2}_{\overline{M_u L_v}} (\overline{L_v} | \lambda_q \sigma_r)$$

$$\tag{1.21}$$

$$= \sum_M (\mu_0 \nu_p | M)(M | \lambda_q \sigma_r). \tag{1.22}$$

The orthonormal RI basis functions do not have the unit cell subscripts because of their delocalized character. After introducing the RI expansion in the transformed integrals we obtain the RI-AO-LT-MP2 analog of Equation (1.11) for the MP2 energy correction

$$E^{\text{MP2}} = \sum_{t=1}^{N_t} (e_t^{SS} - 2e_t^{OS}), \tag{1.23}$$

$$e_t^{OS} = \sum_{\mu\nu\lambda\sigma,\,pqs,\,KL} (\mu_0\overline{\nu_p}|K)(K|\lambda_q\overline{\sigma_s})(\mu_0\nu_p|L)(L|\lambda_q\sigma_s), \tag{1.24}$$

$$e_t^{SS} = \sum_{\mu\nu\lambda\sigma,\,pqs,\,KL} (\mu_0\overline{\nu_p}|K)(K|\lambda_q\overline{\sigma_s})(\mu_0\sigma_s|L)(L|\lambda_q\nu_p), \tag{1.25}$$

where OS and SS are the *opposite-spin* and *same-spin* terms [30]. The main steps of the RI-AO-LT-MP2 algorithm are presented in Scheme 1.2, where N_{RI} is the number of AOs in the RI basis for the whole system. RI bases optimized for MP2 energy calculations usually contain from four to five times more basis functions per unit cell than corresponding regular bases [31]. Scheme 1.2 shows that in the RI-AO-LT-MP2 method the time-limiting step constitutes the contraction of two-electron integrals rather than their transformations.

Generation: $A_{K_s L_v}^{-1/2}$ $[O(N_{\text{RI}}^3)]$
Generation: $(\mu_p \sigma_q | \overline{K_s})$ $[O(N_{\text{RI}} N^2)]$
0th transformation: $(\mu_p v_q | M) = \sum (\mu_p v_q | \overline{K_r}) A_{\overline{K}, M_s}^{-1/2}$ $[O(N_{\text{RI}}^2 N^2)]$
Loop over Laplace points $t = 1, N_t$
    1st transformation: $(\mu_p \delta_u | K) = \sum X_{\mu_p \gamma_v}^t (\gamma_v \delta_u | K)$ $[O(N_{\text{RI}} N^3)]$
    2nd transformation: $(\mu_p \overline{v_q} | K) = \sum Y_{v_q \delta_u}^t (\mu_p \delta_u | K)$ $[O(N_{\text{RI}} N^3)]$
    OS contraction: $e_t^{OS} = \sum (\mu_0 \overline{v_p} | K)(K | \lambda_q \overline{\sigma_s})(\mu_0 v_p | L)(L | \lambda_q \sigma_s)$ $[O(N_{\text{RI}}^2 N^2)]$
    SS contraction: $e_t^{SS} = \sum (\mu_0 \overline{v_p} | K)(K | \lambda_q \overline{\sigma_s})(\mu_0 \sigma_s | L)(L | \lambda_q v_p)$ $[O(N_{\text{RI}} N^3 N_0)]$
End loop over t,

Scheme 1.2

1.3 Implementation details

Our discussion will be oriented toward the RI-AO-LT-MP2 method, because most of the algorithmic features of the AO-LT-MP2 method were implemented in the version including RI. For the sake of simplicity, cell and Laplace point indices will be omitted in cases where notation is obvious.

1.3.1 RI basis extension

In order to proceed along Scheme 1.2 we need to decide the extent to which the RI basis should be replicated. In addition, there is a more fundamental problem of a divergence of the Coulomb metric RI scheme with

PBC [32]. Owing to the structure of AO-LT-MP2 equations, MP2 corrections (Equations [1.12] and [1.13]) can be seen as MP2 expressions for a
large molecule, because all cell indices have finite ranges due to various
decays (see below). Therefore, the RI expansion in Equations (1.12) and
(1.13) is introduced as in the molecular case and is not subjected to PBC.
However, slow decay of fitting coefficients in large systems is still a problem [32]. Several approaches to circumvent this issue have been proposed
recently, e.g., Poisson equation [33], attenuated Coulomb operator [32],
and local domains [34] techniques. Due to the the locality of plain and
transformed AOs, in our implementation we exploit an Ansatz, which re-
sembles the local domain technique [34]. The exact RI representation for
any two-electron integral can be written as

$$(\mu\nu|\lambda\sigma) = (\mu\nu|\overline{K})(\overline{K}|\overline{K})^{-1}(\overline{K}|\lambda\sigma), \tag{1.26}$$

where $\overline{K}$ coincides with either $(\mu\nu|$ or $|\lambda\sigma)$. Therefore, in Equations (1.12)
and (1.13) we use RI functions that overlap with only the bra distributions
of two-electron integrals: $\{\mu_0\nu_p\}$ and $\{\underline{\mu}_0\overline{\nu}_p\}$. The decay of the transformed
and untransformed distributions is exponential with the distance between
the centers of atomic orbitals μ_0 $(\underline{\mu}_0)$ and ν_p $(\overline{\nu}_p)$ [12], and consequently,
the region around the central cell where distributions $\{\mu_0\nu_p\}$ and $\{\underline{\mu}_0\overline{\nu}_p\}$
are nonnegligible is finite. However, the accuracy of the RI approximation
is defined not only by the overlap criteria but also depends on the quality
of the RI basis set that cannot be assessed accurately *a priori*. Thus, in the
current implementation, the number of unit cells for the RI basis replication
(N_c^{RI}) is an external parameter. Given that N_c^{RI} is specified, the $A^{-1/2}$ matrix
is generated by a singular value decomposition.

1.3.2   Basis pair screening

The computational scaling of the integral transformations and contrac-
tions can be reduced by applying various Cauchy–Schwarz-type relations.
For regular and transformed three-index integrals the Cauchy–Schwarz
inequalities are

$$|(\mu\nu|K)| \leq |(\mu\nu|\mu\nu)|^{1/2} = B_{\mu\nu}, \tag{1.27}$$

$$|(\underline{\mu}\nu|K)| \leq |(\underline{\mu}\nu|\underline{\mu}\nu)|^{1/2} \leq \sum_{\sigma} |(\sigma\nu|\sigma\nu)|^{1/2}|X_{\mu\sigma}| = D_{\mu\nu}, \tag{1.28}$$

$$|(\underline{\mu}\overline{\nu}|K)| \leq |(\underline{\mu}\overline{\nu}|\underline{\mu}\overline{\nu})|^{1/2} \leq \sum_{\lambda} |(\mu\nu|\mu\nu)|^{1/2}|Y_{\lambda\nu}| = F_{\mu\nu}. \tag{1.29}$$

The matrices $B_{\mu\nu}$, $D_{\mu\nu}$, and $F_{\mu\nu}$ are used for screening purposes because
their elements have an exponential decay with respect to the distance be-
tween the centers of AOs μ and ν and can be precomputed in negligible

CPU time. However, the matrices $\mathbf{D}$ and $\mathbf{F}$ are usually much denser than the $\mathbf{B}$ matrix, and in practice, it is more beneficial to augment the Cauchy–Schwarz screening of transformed integrals with estimations of the final energy contribution for the particular pair of indices. To illustrate this approach, let us consider the OS part (Equation [1.24]): Even if the $(\mu_0\overline{\nu_p}|K)$ integrals are nonnegligible they are multiplied by their untransformed counterparts $(\mu_0\nu_p|L)$, which decay faster with the distance between μ_0 and ν_p. To obtain the full energy contribution, the SS part also needs to be considered, and in order to treat it efficiently, we introduce atomic pair energies (APEs) defined as the following partial sums:

$$E_{\mu_0\nu_p}(t) = \sum_{\lambda\sigma, rs} T_{\mu_0\lambda_r}^{\nu_p\sigma_s}(t)[2(\mu_0\nu_p|\lambda_r\sigma_s) - (\mu_0\sigma_s|\lambda_r\nu_p)], \tag{1.30}$$

$$E_{\lambda_r\sigma_s}(t) = \sum_{\mu\nu, p} T_{\mu_0\lambda_r}^{\nu_p\sigma_s}(t)[2(\mu_0\nu_p|\lambda_r\sigma_s) - (\mu_0\sigma_s|\lambda_r\nu_p)], \tag{1.31}$$

$$E_{\mu_0\lambda_r}(t) = \sum_{\nu\sigma, ps} T_{\mu_0\lambda_r}^{\nu_p\sigma_s}(t)[2(\mu_0\nu_p|\lambda_r\sigma_s) - (\mu_0\sigma_s|\lambda_r\nu_p)]. \tag{1.32}$$

APEs are convenient for the screening process because they allow us to neglect insignificant pairs of AO indices and have a moderate memory cost. The only problem is obtaining reliable approximations ($\tilde{E}$) for these quantities without introducing significant overhead. In the general case, we use APE values from the previous Laplace point to estimate those for the current Laplace point (see Section 1.3.5), while in the case of $E_{\mu_0\lambda_r}$, a distance relation is employed to connect different APEs for the same Laplace point (see Section 1.3.3).

1.3.3   Distance screening

The main shortcoming of Cauchy–Schwarz screening (Equations [1.27]–[1.29]) is that it does not take into account the decay with respect to the distance (R) between centers of bra and ket distributions in two-electron integrals. This decay is not very useful for screening three-center integrals individually, but it can be efficiently used in the contraction step. Equations (1.24) and (1.25) can be seen as multiplication of regular four-center integrals generated on the fly with their transformed counterparts. In these expressions, distance screening is useful to prescreen weakly interacting distribution pairs $\mu\nu$ and $\lambda\sigma$. One way of approaching this problem is through the multipole moment expansion of two-electron integrals [35]. Following that strategy, Lambrecht et al. proposed rigorous upper bounds for two-electron integrals based on the multipole moment expansion [16, 17]. According to multipole moment consideration, first terms of asymptotic decay for $(\mu\nu|\lambda\sigma)$ and $(\mu\overline{\nu}|\lambda\overline{\sigma})$ are R^{-1} and R^{-3}, respectively [12, 17]. In the AO-LT-MP2 and RI-AO-LT-MP2 algorithms, rigorous upper bounds have not been implemented. Instead, we exploit a simple heuristic scheme that is based on the assumption of monotonic decay for μ-λ pair contributions corresponding to different unit cells

$$E_{\mu_0\lambda_{r+n}}(t) \leq E_{\mu_0\lambda_r}(t), \quad n > 0. \tag{1.33}$$

In the algorithm, $E_{\mu_0\lambda_r}$ pair energies for the rth cell are calculated first, and are used later as $\tilde{E}_{\mu_0\lambda_{r+1}}$ to screen out insignificant $E_{\mu_0\lambda_{r+1}}$ pair energies. From the multipole moment expansion point of view, the assumption of monotonic decay becomes more accurate when the r-cell is far enough from the central cell to avoid a nonnegligible overlap between distributions including μ_0 and λ_r. The performance of our heuristic scheme for the OS and SS APEs

$$E_{\mu_0\lambda_r}^{(OS)}(t) = \sum_{\nu\sigma,\,ps} T_{\mu_0\lambda_r}^{\nu_p\sigma_s}(t)(\mu_0\nu_p|\lambda_r\sigma_s), \tag{1.34}$$

$$E_{\mu_0\lambda_r}^{(SS)}(t) = \sum_{\nu\sigma,\,ps} T_{\mu_0\lambda_r}^{\nu_p\sigma_s}(t)(\mu_0\sigma_s|\lambda_r\nu_p) \tag{1.35}$$

in a *trans*-polyacetylene chain is illustrated in Figures 1.1 and 1.2. This empirical study demonstrates that Equation (1.33) provides an efficient and reliable screening protocol.

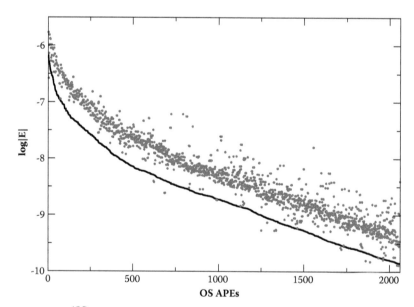

Figure 1.1 $E_{\mu_0\lambda_r}^{(OS)}$ APEs for *trans*-polyacetylene with RI-AO-LT-MP2/cc-pVDZ. Exact APEs are ordered and form the black line, while $\tilde{E}_{\mu_0\lambda_r}^{(OS)}$ estimates are denoted by dots. (From A. F. Izmaylov and G. E. Scuseria, *Phys. Chem. Chem. Phys.* 10, 3421 (2008) by permission of the PCCP Owner Societies.)

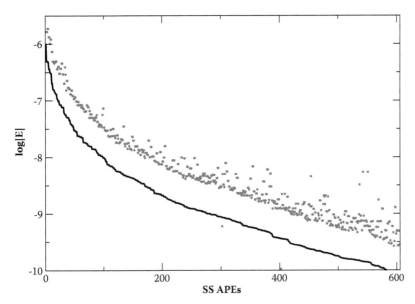

Figure 1.2 $E^{(SS)}_{\mu_0\lambda_r}$ APEs for *trans*-polyacetylene with RI-AO-LT-MP2/cc-pVDZ. Exact APEs are ordered and form the black line, while $\tilde{E}^{(SS)}_{\mu_0\lambda_r}$ estimates are denoted by dots. (From A. F. Izmaylov and G. E. Scuseria, *Phys. Chem. Chem. Phys.* 10, 3421 (2008) by permission of the PCCP Owner Societies.)

1.3.4   Laplace quadratures

In order to obtain a discretized representation of the Laplace integral Equation (1.10), we follow the logarithm transform of the Laplace integration approach introduced by Ayala et al. [13, 36]. In this approach, for a sum E

$$E = \sum_v \frac{z_v}{\Delta_v} = \int_0^\infty dt \sum_v z_v e^{-\Delta_v t}, \qquad (1.36)$$

where Δ_v is positive, a logarithmic change of variable is done first,

$$t = -\frac{1}{\alpha}\ln(x), \qquad (1.37)$$

$$E = -\frac{1}{\alpha}\int_0^1 dx \sum_v z_v x^{(\Delta_v/\alpha)-1}. \qquad (1.38)$$

This change not only produces a polynomic integrand but also makes the integration interval finite. As long as parameter α is smaller or equal to the smallest Δ_v value, the integral Equation (1.38) can be easily integrated numerically using Gauss quadratures. In MP2, α is equal to the double HF band gap for energy calculations and the single HF band gap for band-gap

Table 1.1 Comparison of Fundamental Band Gaps (eV) for Selected Polymers

Method	$(C_2)_\infty$		$(CH_2)_\infty$		$(C_2H_2)_\infty$	
	4-31G	Ref. 39	6-31G*	Ref. 40	6-31G	Ref. 5
HF	6.589	6.60	17.022	17.02	6.187	6.189
MP2/CO		4.50		13.55		4.169
MP2/L3	4.438		13.540		4.248	
MP2/L5	4.473		13.552		4.220	
MP2/L7	4.477		13.555		4.227	

MP2/CO designates available conventional crystal orbital MP2 calculations, whereas MP2/Ln (n=3,5,7) corresponds to the AO-LT-MP2 method using 3, 5, or 7 quadrature points.
Source: P. Y. Ayala, K. N. Kudin, and G. Scuseria, *J. Chem. Phys.* 115, 9698 (2001).

calculations. Generally, 5 to 7 (3 to 5) quadrature points are required to achieve μE_h (10 meV) accuracy in energy (band-gap) calculations. Some illustrations of the accuracy of the logarithm transform scheme for typical one-dimensional systems are given in Table 1.1. Logarithm transformations have proven to be valuable in numerical integration of exponential functions in other areas as well, including radial quadratures in density-functional [37] and effective core potential calculations [38].

An alternative approach to the Laplace integral evaluation is a least square fit of $1/x$ over the range of denominator values

$$\frac{1}{x} \approx \sum_{t=1}^{N_t} w_t e^{-\alpha_t x}. \tag{1.39}$$

This approach was first formulated by Häser and Almlöf [41], and later used and explored in many other studies [12,42,43]. Although [42] and [43] extensively compare different least square fitting approaches, we are not aware of a systematic study that emphasizes a superiority of the least square fit approach over the one using the logarithm transformation. According to data presented in [42] and [43], in order to achieve μE_h energy accuracy, best least square fitting techniques require the same number of quadrature points (5–7) as the logarithm transformation approach.

1.3.5   Relation between quadrature points

Due to the discretized form of the Laplace integral, in AO-LT-MP2, all properties of interest are weighted sums of quadrature point contributions. The algebraic structure of these sums (Equations [1.11]–[1.12]) is identical for all quadrature points, but the sums involve different weighted density matrices for each quadrature point. To increase efficiency of the algorithm it is desirable to obtain a relation between APEs at different quadrature points. The $E_{\mu\nu}$ pair energy can be considered as a function of t

$$E_{\mu\nu}(t) = \sum_{\lambda\sigma} T^{\nu\sigma}_{\mu\lambda}(t)[2(\mu\nu|\lambda\sigma) - (\mu\sigma|\lambda\nu)]. \tag{1.40}$$

To elucidate an important functional dependence, it is useful to rewrite $E_{\mu\nu}$ as

$$E_{\mu\nu}(t) = \sum_{iajb} B^{\mu\nu}_{iajb} e^{\Delta_{iajb} t}, \tag{1.41}$$

where $B^{\mu\nu}_{iajb}$ are the elements of some tensor, and $\Delta_{iajb} = \epsilon_i + \epsilon_j - \epsilon_a - \epsilon_b$ are negative energy denominators. If all elements of $B^{\mu\nu}_{iajb}$ had the same sign for the particular $\mu\nu$ pair, the function $E_{\mu\nu}(t)$ would be monotonic with t, and estimation of $E_{\mu\nu}(t')$ from $E_{\mu\nu}(t)$ would be straightforward. Unfortunately, in general, the $B^{\mu\nu}_{iajb}$ elements have different signs, and $E_{\mu\nu}(t)$ is not a monotonic function. However, the $E_{\mu\nu}(t)$ function can be split into two monotonic parts by grouping all elements of the same sign

$$E_{\mu\nu}(t) = \sum_{\{iajb\}}^{-} B^{\mu\nu}_{iajb} e^{\Delta_{iajb} t} + \sum_{\{iajb\}}^{+} B^{\mu\nu}_{iajb} e^{\Delta_{iajb} t} \tag{1.42}$$

$$= E^{-}_{\mu\nu}(t) + E^{+}_{\mu\nu}(t), \tag{1.43}$$

where $\sum^{-}$ and $\sum^{+}$ are the sums over all negative and positive elements of B^{ab}_{ij}, respectively. This splitting allows us to put bounds on a value of $E_{\mu\nu}(t)$ for $t_2 = t_1 + \delta t$

$$E_{\mu\nu}(t_2) \leq E^{+}_{\mu\nu}(t_1) e^{\Delta_{min}\delta t} + E^{-}_{\mu\nu}(t_1) e^{\Delta_{max}\delta t}, \tag{1.44}$$

and

$$E_{\mu\nu}(t_2) \geq E^{-}_{\mu\nu}(t_1) e^{\Delta_{min}\delta t} + E^{+}_{\mu\nu}(t_1) e^{\Delta_{max}\delta t}, \tag{1.45}$$

where $\Delta_{min} = [\min_{ij,ab} \Delta_{iajb}]$ and $\Delta_{max} = [\max_{ij,ab} \Delta_{iajb}]$. To evaluate these bounds for an arbitrary $t_3 = t_1 + \tilde{\delta}t$, $E^{+}_{\mu\nu}(t_1)$ and $E^{-}_{\mu\nu}(t_1)$ need to be available. Reversing the expressions in Equations (1.44) and (1.45) gives rise to the following bounds on $E^{\pm}_{\mu\nu}(t_1)$

$$E^{\pm}_{min} \leq E^{\pm}_{\mu\nu}(t_1) \leq E^{\pm}_{max}, \tag{1.46}$$

where

$$E^{-}_{min} = \frac{E_{\mu\nu}(t_2) - E_{\mu\nu}(t_1) e^{\Delta_{max}\delta t}}{e^{\Delta_{min}\delta t} - e^{\Delta_{max}\delta t}}, \tag{1.47}$$

$$E^{-}_{max} = \min\left[0, \frac{E_{\mu\nu}(t_2) - E_{\mu\nu}(t_1) e^{\Delta_{min}\delta t}}{e^{\Delta_{max}\delta t} - e^{\Delta_{min}\delta t}}\right], \tag{1.48}$$

$$E^{+}_{min} = \max\left[0, E_{\mu\nu}(t_1) - E^{-}_{max}\right], \tag{1.49}$$

$$E^{+}_{max} = \max\left[0, E_{\mu\nu}(t_1) - E^{-}_{min}\right]. \tag{1.50}$$

Thus, given that $E_{\mu\nu}(t_1)$ and $E_{\mu\nu}(t_2)$ are calculated, an estimate for $E_{\mu\nu}(t_3)$ is

$$E_{\mu\nu}(t_3) \leq E_{max}^{+}e^{\Delta_{min}\tilde{\delta}t} + E_{max}^{-}e^{\Delta_{max}\tilde{\delta}t}, \tag{1.51}$$

and

$$E_{\mu\nu}(t_3) \geq E_{min}^{-}e^{\Delta_{min}\tilde{\delta}t} + E_{min}^{+}e^{\Delta_{max}\tilde{\delta}t}. \tag{1.52}$$

It is worth emphasizing that any property or partially contracted sum for a particular Laplace point can be bound with the derived inequalities.

1.3.6 Transformation and contraction algorithms

For the sake of simplicity, we will skip the disk operations involved in the actual implementation and assume that all quantities can be fit in the core memory. In preparation for the zeroth integral transformation, we generate three-center integrals so that the fastest index [44] corresponds to the RI basis $(\overline{K}|[\mu\nu])$, and AO basis pairs contain only nonnegligible $\mu\nu$ distributions. A regular BLAS library matrix-matrix multiplication procedure is the most effective way of transforming $(\overline{K}|[\mu\nu])$ integrals because of the dense character of the $A_{KL}^{-1/2}$ matrix. In contrast, it is more efficient to exploit a partial compressed matrix multiplication in the first transformation (Scheme 1.2) because in this case, AO indices are transformed and can be subjected to various screenings. First, integrals after the zeroth transformation $(K|[\mu\nu])$ are transposed to a set of matrices $(K|[\nu]_\mu, \mu)$. The matrix notation $(K|[\nu]_\mu, \mu)$ specifies the part of $(K|\mu\nu)$ integrals with a fixed μ basis function and ν indices restricted to a set of $[\nu]_\mu = 1, \ldots, N_\mu$ basis functions that overlap with μ ($B_{\mu\nu} >$ Threshold). Then, each $(K|[\nu]_\mu, \mu)$ matrix is compressed on the fly by using the standard compressed format [45] (see Scheme 1.3). The compression procedure gives rise to three arrays per $(K|[\nu]_\mu, \mu)$ matrix: the IA array denotes the beginning and the end of compressed $[K]$ indices corresponding to a $[\nu]_\mu$ element, JA contains a mapping between compressed $[K]$ and real K indices, and the CA array stores values of significant integrals.

```
Loop over μ, 1 ≤ μ ≤ N
        Compress (K|[ν]_μ, μ) into CA, IA, and JA arrays
        Loop over [ν]_μ, 1 ≤ [ν]_μ ≤ N_μ
                [K]₁ = IA([ν]_μ)
                [K]_END = IA([ν]_μ + 1) - 1
                If (Ẽ_μν > Threshold ) then
                        Loop over λ, 1 ≤ λ ≤ N
                                If (Min(Ẽ_λ, X_λμ, D_λμ) > Threshold) then
                                        Loop over [K], [K]₁ ≤ [K] ≤ [K]_END
                                                K = JA([K])
                                                (K|λν) = (K|λν) + X_λμCA([K])
```

```
                        End loop over [K]
                    EndIf
                End loop over λ
            EndIf
        End loop over [v]_μ
    End loop over μ
```

Scheme 1.3

We do not compress the matrices **X**, **Y**, **W**, and **Z** because they are not usually very sparse. The final result of each transformation is also gathered in the uncompressed form. In order to reduce the number of elements treated in the internal loops and to avoid cache memory reading faults [46], we compress integrals before their transformation. These compressions are done outside the internal loops and thus do not create any significant overhead. We perform the second transformation in a very similar way to the first one with commensurate substitution of weighted density matrices, integrals, and Cauchy–Schwarz screening matrices. In Scheme 1.3, single orbital estimates $\tilde{E}_\lambda = \sum_\mu \tilde{E}_{\mu\lambda}$ and APEs $\tilde{E}_{\mu\nu}$ are employed only for later Laplace points, when reliable estimates are evaluated. In the contractions of Equations (1.24) and (1.25), only AO indices of three-center integrals are compressed, while the RI indices are left uncompressed (see Scheme 1.4). This partial compression is more effective than the full one because $(K|[v]_\mu, \mu)$ matrices or their transformed analogs are not sparse enough for the sparse matrix multiplication $([v]_\mu, \mu|[\lambda]_\sigma, \sigma) = (K|[\lambda]_\sigma, \sigma)(K|[v]_\mu, \mu)$ to be faster than a standard BLAS routine.

```
Loop over μ, 1 ≤ μ ≤ N
        Read (K|μv) into (K|[v]_μ, μ)
        Read (K|μv̄) into (K|[v̄]_μ, μ)
        Loop over λ, 1 ≤ λ ≤ N
            If (Ẽ_μλ > Threshold) then
                    Read (K|λσ) into (K|[σ]_λ, λ)
                    Matrix multiplications ([v]_μ, μ|[σ]_λ, λ) = (K|[v]_μ, μ)(K|[σ]_λ, λ)
                    If (Ẽ_μλ^(OS) > Threshold) then
                        Read (K|λσ̄) into (K|[σ̄]_λ, λ)
                        Matrix multiplication ([v̄]_μ, μ|[σ̄]_λ, λ) = (K|[v̄]_μ, μ)(K|[σ̄]_λ, λ)
                        Dot product E_μλ^(OS) = ([v]_μ, μ|[σ]_λ, λ)([v̄]_μ, μ|[σ̄]_λ, λ)
                    EndIf
                    If (Ẽ_μλ^(SS) > Threshold) then
                        Read (K|μσ̄) into (K|[σ̄]_λ, μ)
                        Read (K|λv̄) into (K|[v̄]_μ, λ)
                        Matrix multiplication ([v̄]_μ, λ|[σ̄]_λ, μ) = (K|[v̄]_μ, λ)(K|[σ̄]_λ, μ)
                        Dot product E_μλ^(SS) = ([v]_μ, μ|[σ]_λ, λ)([v̄]_μ, λ|[σ̄]_λ, μ)
                    EndIf
            EndIf
        End loop over λ
End loop over μ
```

Scheme 1.4

1.3.7   Lattice summations

Formally, the number of unit cells in the real space representation is equal to the number of k-points for the Brillouin zone sampling. On the other hand, convergence characteristics of different quantities with the number of cells are not the same, and it is more efficient to use different cell ranges. Equations (1.24) and (1.25) have three independent cell indices p, q, and s. The Cauchy–Schwarz inequality restricts p and s cells to be in the vicinity of 0th and r^{th} cells. The $|0 - r|^{-6}$ ($|0 - r|^{-5}$) distance decay of the MP2 energy (band gap) defines a finite range for the r index [13]. However, prior to the contraction step in Equations (1.24) and (1.25), the untransformed integrals are used in the transformation procedure

$$(K|\underline{\lambda}_r \overline{\sigma}_s) = \sum_{\mu\nu,\nu u} X^t_{\mu_\nu \lambda_r} Y^t_{\nu_u \sigma_s} (K|\mu_\nu \nu_u). \tag{1.53}$$

Here, indices v and u have a larger spatial extension than indices r and s because of the coupling through the X and Y matrices. Although decay properties of these matrices can be rigorously obtained without large overhead, estimates they provide for the size of the spatial framework N_c^{max} are usually too conservative. Thus, in our implementation, we do not use an *a priori* method to define the range of the r index (r_{max}^{MP2}) in the contraction step and the spatial framework size N_c^{max}, but rather consider them as external parameters in our calculations.

1.3.8   Symmetry

In the AO-LT-MP2 and RI-AO-LT-MP2 implementations, crystal symmetry has not been fully exploited, but it is employed in the most time-consuming parts. We use symmetry by introducing symmetry equivalent atomic pairs defined as pairs of atoms that can be transformed into each other by operations of the point group of the unit cell. This notion becomes useful after splitting the AO-LT-MP2 energy in atomic pair contributions [47]

$$E^{MP2} = \sum_{A,B} E^{MP2}_{AB}, \tag{1.54}$$

where A and B are all pairs of atoms with corresponding energy contributions defined as

$$E^{MP2}_{AB} = \sum_{\mu_0 \in A, \lambda_r \in B} E_{\mu_0 \lambda_r}. \tag{1.55}$$

Thus, to speed up the contraction step (Scheme 1.4) we generate $E_{\mu_0\lambda_r}$ APEs only over symmetry nonequivalent atomic pairs and scale them appropriately to account for the symmetry equivalent counterparts.

1.4 Benchmark calculations

The purpose of this section is twofold: (1) to illustrate improvements in computational efficiency related to the RI approximation and the accuracy level that one could expect in the RI-AO-LT-MP2 method, and (2) to review some applications of the AO-LT-MP2 method to evaluation of the fundamental band gap in various periodic systems. The AO-LT-MP2 and RI-AO-LT-MP2 methods have been implemented in the development version of the GAUSSIAN program [48]. Details of our PBC HF implementation can be found in [18], [49], and [50].

1.4.1 RI approximation

For a comparative assessment of the RI-AO-LT-MP2 and AO-LT-MP2 methods, we have computed unit cell energies and band gaps of *trans*-polyacetylene (tPA) and *anti*-transoid polymethineimine (tPMI), which are depicted in Figure 1.3. All calculations for these systems are done using 10^{-8} threshold in Schemes 1.3 and 1.4, 1024 k-points for the Brillouin zone sampling, and five-point quadratures in the Laplace transformation. We use the cc-pVDZ basis set with its RI counterpart optimized [31] for molecular RI-MP2 calculations. Since the MP2 gradients with PBC have not been implemented yet, all geometries were optimized with the Perdew-Berke-Ernzerhof hybrid [51] (PBEh) functional. Previous studies [52, 53] reveal

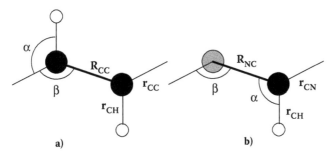

Figure 1.3 PBEh/cc-pVDZ optimized structures of (A) *trans*-polyacetylene [$\alpha = 117.25°$, $\beta = 124.32°$, $R_{CC} = 1.3669$ Å, $r_{CH} = 1.0966$ Å, $r_{CC} = 1.4241$ Å, lattice constant $= 2.4681$ Å] and (B) *anti*-transoid polymethineimine [$\alpha = 120.92°$, $\beta = 118.15°$, $R_{NC} = 1.3242$ Å, $r_{CH} = 1.1176$ Å, $r_{CN} = 1.3241$ Å, lattice constant $= 2.2718$ Å]. (From A. F. Izmaylov and G. E. Scuseria, *Phys. Chem. Chem. Phys.* 10, 3421 (2008) by permission of the PCCP Owner Societies.)

that PBEh gives tPA geometry and band gap in very good agreement with
experimental values and other trustworthy theoretical studies [13,54,55].
tPMI is isoelectronic to tPA, and therefore one could expect good PBEh
performance for tPMI and similarity in structural features for both systems. However, in contrast to the tPA case, our periodic PBEh/cc-pVDZ
and PBEh/6-31G(d) geometry optimizations show no bond-length alternation for tPMI [56]. The contradiction of this result to the outcome of the
PBEh oligomeric calculations in [57] suggests that the discrepancy originates from a periodicity constraint in our PBC study. After relaxing this
constraint by taking a super cell with 16 CHN fragments as a new unit cell,
we have observed a non-negligible bond-length alternation. The geometry
obtained does not allow a reduction of the 16-CHN-fragment unit cell to a
unit cell with a smaller number of CHN fragments. Although the PBEh/
cc-pVDZ method with the minimal unit cell (one CHN fragment) does not
capture the bond-length alternation feature, the geometry it produces is
still suitable for benchmark purposes.

First, we would like to address the convergence of the AO-LT-MP2
method with the maximum range of r index (r_{max}^{MP2}) in Equations (1.11)–
(1.12) and real-space cell framework extension (N_c^{max}). Two series of calculations testing this convergence are presented in Table 1.2. Results
with $r_{max}^{MP2} = 21$ and $N_c^{max} = 29$ are considered to be converged within
μE_h and meV accuracy in energy and band gap corrections (Table 1.3).
Therefore, for the presented polymers, parameter values $r_{max}^{MP2} = 13$ and
$N_c^{max} = 19$ minimize computational efforts and still provide acceptable
accuracy.

Table 1.2 Differences in Energy (ΔE, μE_h) and Band Gap
(ΔE_g, eV) MP2 Corrections in AO-LT-MP2/cc-pVDZ
Calculations of *Trans*-Polyacetylene (tPA) and *Anti*-transoid
Polymethineimine (tPMI) as a Function of r_{max}^{MP2} and N_c^{max}

$r_{max}^{MP2} / N_c^{max}$	ΔE		ΔE_g	
	tPA	tPMI	tPA	tPMI
11/17	5.5	11.7	0.024	0.040
11/19	5.4	12.6	0.024	0.040
11/21	5.6	12.3	0.024	0.040
13/17	3.5	5.0	0.015	0.023
13/19	3.4	5.8	0.015	0.023
13/21	3.6	5.7	0.015	0.023
17/25	1.0	0.9	0.005	0.006
19/27	0.5	0.0	0.002	0.002

Source: A. F. Izmaylov and G. E. Scuseria, *Phys. Chem. Chem. Phys.* 10,
3421 (2008) by permission of the PCCP Owner Societies.

Table 1.3 Energies (μE_h) and Band Gaps (E_g, eV) in
HF/cc-pVDZ and Corresponding AO-LT-MP2/cc-pVDZ
Corrections for *Trans*-Polyacetylene (tPA) and *Anti*-Transoid
Polymethineimine (tPMI) with $r_{max}^{MP2} = 21$ and $N_c^{max} = 29$

	tPA	tPMI
E^{HF}	−76.894 360	−92.890 069
E^{MP2}	−0.271 668	−0.293 876
E_g^{HF}	5.823	7.421
E_g^{MP2}	−2.406	−4.094

Source: A. F. Izmaylov and G. E. Scuseria, *Phys. Chem. Chem. Phys.*
10, 3421 (2008) by permission of the PCCP Owner Societies.

Our implementation of the RI approximation for periodic systems requires defining the cell range for the RI basis replication. Table 1.4 illustrates deviations of the RI-AO-LT-MP2 energy and band gap corrections from those of AO-LT-MP2 for our test systems. It is worth noting that the RI errors for the energy calculations in the tPA case agree well with an estimate that can be done from molecular RI-MP2 calculations. The RI errors at the RI-MP2/cc-pVDZ level of theory for butadiene and hexatriene are 102 μE_h and 148 μE_h [10], respectively. Hence, we can deduce that the average C_2H_2 unit error is approximately 50 μE_h, and we can expect even larger errors in the periodic case due to the longer intercell interaction range. According to Table 1.4, N_c^{RI} can be chosen as low as 7 cells for all cases.

To conclude our RI-AO-LT-MP2 assessment, we present a comparison of CPU timings for the RI-AO-LT-MP2 and AO-LT-MP2 algorithms.

Table 1.4 *Trans*-Polyacetylene (tPA) and *Anti*-transoid
Polymethineimine (tPMI) RI-AO-LT-MP2/cc-pVDZ Energy (ΔE_0,
μE_h) and Band Gap (ΔE_g, meV) Deviations from
AO-LT-MP2/cc-pVDZ Values with $r_{max}^{MP2} = 11$ and $N_c^{max} = 19$ as a
Function of the RI Basis Cell Range (N_c^{RI})

N_c^{RI}	ΔE		ΔE_g	
	tPA	tPMI	tPA	tPMI
7	−59.2	−29.3	−0.021	−0.006
9	−56.7	−3.9	−0.015	−0.005
11	−69.0	3.2	−0.014	−0.008
13	−92.5	2.9	−0.010	−0.005
15	−105.9	0.4	−0.008	−0.003
17	−70.0			
19	−70.4			

Source: A. F. Izmaylov and G. E. Scuseria, *Phys. Chem. Chem. Phys.* 10, 3421
(2008) by permission of the PCCP Owner Societies.

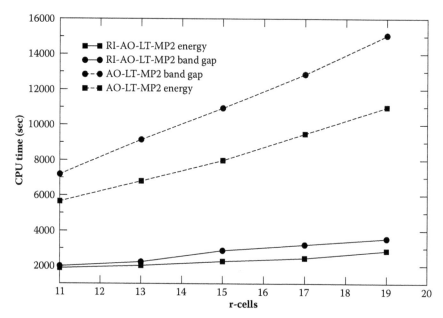

Figure 1.4 Power5 CPU time for *trans*-polyacetylene MP2 unit cell energy and band gap correction calculations with the RI-AO-LT-MP2/cc-pVDZ and AO-LT-MP2/cc-pVDZ methods as a function of $r_{\text{max}}^{\text{MP2}}$. (From A. F. Izmaylov and G. E. Scuseria, *Phys. Chem. Chem. Phys.* 10, 3421 (2008) by permission of the PCCP Owner Societies.)

In addition to parameters that are shared by both algorithms (e.g., $r_{\text{max}}^{\text{MP2}}$ and N_c^{max}), the RI-AO-LT-MP2 algorithm performance also depends on the extension of the fitting domain (N_c^{RI}). The CPU time comparisons for tPA (Figure 1.4) and tPMI (Figure 1.5) are done with variable $r_{\text{max}}^{\text{MP2}}$, fixed $N_c^{\text{RI}} = 7$, and $N_c^{\text{max}} = r_{\text{max}}^{\text{MP2}} + 8$ values. These values are chosen on the basis of previous calculations, so that they can guarantee a satisfactory accuracy (see Tables 1.2 and 1.4). Figures 1.4 and 1.5 show that introducing the RI expansion leads to a significant CPU time reduction for the test systems.

In general, one might need to increase the parameter N_c^{RI} as well. In this case, the generation of the $A_{K_s L_t}^{-1/2}$ matrix and the zeroth transformation of two-electron integrals could start to contribute substantially to the total CPU time and even dominate over the rest of the RI-AO-LT-MP2 algorithm. However, in practice, the value of N_c^{RI} needed is always less than $r_{\text{max}}^{\text{MP2}}$, because the overall decay is exponential with N_c^{RI} and only polynomial with $r_{\text{max}}^{\text{MP2}}$. Therefore, it is always possible to improve the efficiency of the AO-LT-MP2 method by using the RI technique and to avoid substantial loss of accuracy.

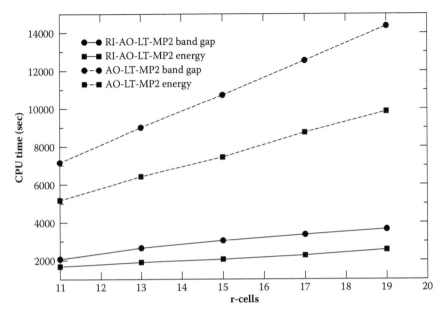

Figure 1.5 Power5 CPU time for *anti*-transoid polymethineimine MP2 unit cell energy and band gap correction calculations with the RI-AO-LT-MP2/cc-pVDZ and AO-LT-MP2/cc-pVDZ methods as a function of r_{max}^{MP2}. (From A. F. Izmaylov and G. E. Scuseria, *Phys. Chem. Chem. Phys.* 10, 3421 (2008) by permission of the PCCP Owner Societies.)

1.4.2   AO-LT-MP2 applications

The RI-AO-LT-MP2 method for periodic systems has become available relatively recently [22], and therefore, not many applications have been done with it yet. On the other hand, the AO-LT-MP2 method was implemented several years ago and was successfully applied to various periodic systems: tPA single and multiple chains, [13, 52] a tPA two-dimensional monolayer [52], a polyphenylenevinylene (PPV) chain [13], and a two-dimensional BN sheet [13]. For all these systems the MP2 fundamental band gap can be compared directly with experiment, and thus, can serve to assess the accuracy of the method. Usually, HF dramatically overestimates experimental band gaps, and the MP2 correction, although it reduces band gaps, cannot fully compensate for the initial HF overestimation (see Table 1.5). There are three main reasons for the discrepancy between the MP2 and experimental band gaps in Table 1.5: (1) incomplete basis sets, this is especially true for the 2D BN case, (2) the MP2 correction represents only a part of the total electron correlation energy, and (3) experimental estimates have been obtained on the condensed (thin film or bulk) material, while theoretical calculations performed on idealized isolated infinite

Table 1.5 The MP2 Band Gap for Various Systems

System	HF	MP2	Experiment
tPA chain	6.1	3.68	2 [63]
PPV chain	7.2	4.65	2.88 ± 0.14 [64]
2D BN sheet	13.7	7.5	5.8 [65]

Values were calculated with the AO-LT-MP2 method and the 6-31G* basis set, except the BN sheet case where the STO-3G Basis was employed. The description of geometrical configurations and other details of calculations can be found in [13].

Source: P. Y. Ayala, K. N. Kudin, and G. Scuseria, *J. Chem. Phys.* 115, 9698 (2001).

one- or two-dimensional structures. Possible variations of MP2 band gaps related to use of different basis sets are illustrated in Table 1.6 for the single tPA chain optimized at the MP2/6-31G* level of theory [52]. Results in Table 1.6 indicate that for achieving meV convergence, one needs to employ basis sets larger than the cc-pVTZ one. The RI approximation can be of great help in extending basis set size limits, and work in this direction is underway. One of the ways to extend MP2 band gap expression (Equations [1.5]–[1.8]), without going explicitly to higher orders of perturbation theory (PT), is connecting the MP2 band gap with the diagonal Dyson quasiparticle energy difference $\epsilon_{LUCO}^{DD}(\mathbf{k}_{min}) - \epsilon_{HOCO}^{DD}(\mathbf{k}_{min})$, where

$$\epsilon_g^{DD}(\mathbf{k}_{min}) = \epsilon_g^{HF}(\mathbf{k}_{min}) + \Sigma_{gg}(\epsilon_g^{DD}(\mathbf{k}_{min})), \qquad (1.56)$$

and $\Sigma_{gg}(E)$ is the diagonal element of a self-energy matrix. It can be shown that the MP2 expression is the result of the first iteration of the diagonal Dyson correction (Equation [1.56]) with the self-energy matrix obtained at the second-order PT [58]. Thus, further iterations of Equation 1.56 can

Table 1.6 The HF and MP2 Band Gaps of
the Single Chain of *Trans*-Polyacetylene
with Various Basis Sets

Basis	HF	MP2
STO-3G	6.61	5.42
3-21G	5.80	3.91
6-31G	5.79	3.85
6-31G*	5.96	3.67
6-31G**	5.95	3.66
6-311G**	5.93	3.54
cc-pVDZ	5.94	3.55
cc-pVTZ	5.93	3.37

Source: R. Pino and G. E. Scuseria, *J. Chem. Phys.*
121, 8113 (2004).

Table 1.7 The MP2 and Diagonal Dyson Band
Gaps of the Single Chain of
Trans-Polyacetylene with Various Basis Sets

Basis	MP2	Dyson
6-31G	3.85	4.14
6-31G*	3.67	4.01
6-31G**	3.66	4.01
6-311G**	3.54	3.91

Source: R. Pino and G. E. Scuseria, *J. Chem. Phys.* 121, 2553 (2004).

be done in an attempt to improve the MP2 band gap results. Moreover, the iterative solution of the diagonal Dyson equation with the Laplace transform does not require additional manipulations with two-electron integrals, and therefore, it has the same computational cost as the AO-LT-MP2 band gap calculation [58]. Unfortunately, the iterative correction does not improve MP2 results, but rather worsens them by reducing the magnitude of the MP2 corrections, thus letting the HF overestimation prevail (see Table 1.7). Another approach to improving the amount of electron correlation obtained within the second order of PT is to substitute the zeroth order HF Hamiltonian with its DFT counterpart. Generally, semilocal DFT captures the electronic structure of solids better than HF. The GW approximation, which is closely related to the diagonal Dyson approximation but employs semilocal DFT orbitals and orbital energies, performs very well for band gaps in general [59] and for tPA (2.1 eV) and PPV (3.3 eV) systems in particular [60]. Thus, a recent interest in incorporating the MP2 correlation energy expression into the KS-DFT framework of so-called *double-hybrid* density functionals [61, 62] can be also motivated by the success of the GW approach in fundamental band gap calculations. As for the third shortcoming of initial MP2 studies of quasi-one-dimensional polymers [13], the influence of the chain–chain interaction for tPA chains has also been investigated with the AO-LT-MP2 method by Pino et al. [52] (Table 1.8). Increasing the number of chains gives rise to a growth of the binding energy per C_2H_2 fragment, and a shift of the MP2 band gaps in the right direction.

1.5 Conclusion

We have reviewed the AO-LT-MP2 and RI-AO-LT-MP2 methods for calculation of the unit cell energy and fundamental band gap in periodic systems. The AO Laplace transform formulation allows one to avoid computationally expensive multidimensional k-integration and to employ various screening schemes. Introducing the RI expansion significantly reduces the

Table 1.8 The MP2 Minimal Band Gap and Binding Energy per C_2H_2 Unit for Various Numbers of *Trans*-Polyacetylene Chains Calculated with the AO-LT-MP2 Method and the 6-31G** Basis Set

System	Minimal Band Gap,[a] eV	Binding Energy, kcal/mol
Single chain	3.67	—
Double chain	2.98	1.1
Triple chain	1.95	2.0
2D sheet	2.59	2.2

The description of geometrical configurations and other details of calculations can be found in [52].

[a] Band gaps are direct for all chains and indirect for the 2D sheet.

Source: R. Pino and G. E. Scuseria, *J. Chem. Phys.* 121, 8113 (2004).

prefactor of the two-electron integral transformation, but it also increases the formal scaling of the contraction step. In order to improve the overall computational efficiency, the RI-AO-LT-MP2 algorithm combines various screening protocols with standard BLAS matrix multiplication routines. The RI-AO-LT-MP2 formalism allows us to restrict the RI fitting domain to the vicinity of the central unit cell without substantial loss of accuracy. This restriction not only makes the RI expansion computationally inexpensive but also alleviates the problem of the Coulomb metric RI fitting divergence with PBC. For the one-dimensional examples considered here, it was found that the RI approximation introduces errors comparable with those in molecular calculations, and the RI-AO-LT-MP2 method works 4 to 5 times faster than the AO-LT-MP2 method with 100 μE_h and 10 meV accuracy in energy and band gap, respectively. The comparison of MP2 fundamental band gaps with experimental results suggests that to obtain good agreement, one needs to use large basis sets and to include non-covalently bound components of the system. We believe that further improvements of calculated band gaps in solids can be achieved through developing perturbation theories built on top of the DFT zeroth order approximation.

Acknowledgments

Portions of this work have been supported by the U.S. National Science Foundation, the U.S. Department of Energy, and the Welch Foundation.

References

[1] S. Tsuzuki and H. P. Lüthi, "Interaction energies of van der Waals and hydrogen bonded systems calculated using density functional theory: Assessing the PW91 model," *J. Chem. Phys.* **114**, 3949 (2001).

[2] B. J. Lynch and D. G. Truhlar, "How well can hybrid density functional methods predict transition state geometries and barrier heights?" *J. Phys. Chem. A* **105**, 2936 (2001).

[3] J. Gauss, "Effects of electron correlation in the calculation of nuclear magnetic resonance chemical shifts," *J. Chem. Phys.* **99**, 3629 (1993).

[4] S. Suhai, "Quasiparticle energy-band structures in semiconducting polymers: Correlation effects on the band gap in polyacetylene," *Phys. Rev. B* **27**, 3506 (1983).

[5] J. Sun and R. J. Bartlett, "Second-order many-body perturbation-theory calculations in extended systems," *J. Chem. Phys.* **104**, 8553 (1996).

[6] J. A. Pople, J. S. Binkley, and R. Seeger, "Theoretical models incorporating electron correlation," *Int. J. Quantum Chem. S* **10**, 1 (1976).

[7] P. Pulay, "Localizability of dynamic electron correlation," *Chem. Phys. Lett.* **100**, 151 (1983).

[8] S. Saebø and P. Pulay, "Fourth-order Møller–Plessett perturbation theory in the local correlation treatment. I. Method," *J. Chem. Phys.* **86**, 914 (1987).

[9] S. Saebø and P. Pulay, "A low-scaling method for second order Møller–Plesset calculations," *J. Chem. Phys.* **115**, 3975 (2001).

[10] H. J. Werner, F. R. Manby, and P. J. Knowles, "Fast linear scaling second-order Møller–Plesset perturbation theory (MP2) using local and density fitting approximations," *J. Chem. Phys.* **118**, 8149 (2003).

[11] M. Häser, "Møller–Plesset (MP2) perturbation theory for large molecules," *Theor. Chim. Acta* **87**, 147 (1993).

[12] P. Y. Ayala and G. Scuseria, "Linear scaling second-order Møller–Plesset theory in the atomic orbital basis for large molecular systems," *J. Chem. Phys.* **110**, 3660 (1999).

[13] P. Y. Ayala, K. N. Kudin, and G. Scuseria, "Atomic orbital Laplace-transformed second-order Møller–Plesset theory for periodic systems," *J. Chem. Phys.* **115**, 9698 (2001).

[14] C. Pisani, M. Busso, G. Capecchi, S. Casassa, R. Dovesi, L. Maschio, C. Zicovich-Wilson, and M. Schütz, "Local-MP2 electron correlation method for nonconducting crystals," *J. Chem. Phys.* **122**, 094113 (2005).

[15] D. L. Strout and G. E. Scuseria, "A quantitative study of the scaling properties of the Hartree–Fock method," *J. Chem. Phys.* **102**, 8448 (1995).

[16] D. S. Lambrecht, B. Doser, and C. Ochsenfeld, "Multipole-based integral estimates for the rigorous description of distance dependence in two-electron integrals," *J. Chem. Phys.* **123**, 184101 (2005).

[17] D. S. Lambrecht and C. Ochsenfeld, "Rigorous integral screening for electron correlation methods," *J. Chem. Phys.* **123**, 184102 (2005).

[18] A. F. Izmaylov, G. E. Scuseria, and M. J. Frisch, "Efficient evaluation of short-range Hartree–Fock exchange in large molecules and periodic systems," *J. Chem. Phys.* **125**, 104103 (2006).

[19] B. Dunlap, J. Connolly, and J. Sabin *J. Chem. Phys.* **71**, 4993 (1979).

[20] L. Maschio, D. Usvyat, F. R. Manby, S. Casassa, C. Pisani, and M. Schütz, "Fast local-MP2 method with density-fitting for crystals. I. Theory and algorithms," *Phys. Rev. B* **76**, 075101 (2007).

[21] D. Usvyat, L. Maschio, F. R. Manby, S. Casassa, M. Schütz, and C. Pisani, "Fast local-MP2 method with density-fitting for crystals. II. Test calculations and application to the carbon dioxide crystal," *Phys. Rev. B* **76**, 075102 (2007).

[22] A. F. Izmaylov and G. E. Scuseria, "Resolution of the identity atomic orbital Laplace transformed second order Møller–Plesset theory for nonconducting periodic systems," *Phys. Chem. Chem. Phys.* **10**, 3421 (2008).

[23] N. W. Ashcroft and N. D. Mermin, *Solid State Physics*, Saunders Colege, Orlando, Florida, 1976.

[24] A. Szabo and N. S. Ostlund, *Modern Quantum Chemistry*, Macmillan, New York, 1982.

[25] J. J. Ladik, *Quantum Theory of Polymers as Solids*, Plenum, New York, 1988.

[26] G. Onida, L. Reining, and A. Rubio, "Electronic excitations: density-functional versus many-body Green's-function approaches," *Rev. Mod. Phys.* **74**, 601–659 (2002).

[27] J. Almlöf, "Elimination of energy denominators in Møller–Plesset perturbation theory by a Laplace transform approach," *Chem. Phys. Lett.* **181**, 319 (1991).

[28] N_t is usually 3-5 for band gap calculations and 5-7 for energy calculations.

[29] The rectangular quadrature is used for all transformations between real and reciprocal spaces.

[30] Y. Jung, R. C. Lochan, A. D. Dutoi, and M. Head-Gordon, "Scaled opposite-spin second order Møller–Plesset correlation energy: An economical electronic structure method," *J. Chem. Phys.* **121**, 9793 (2004).

[31] F. Weigend, A. Köhn, and C. Hättig, "Efficient use of the correlation consistent basis sets in resolution of the identity MP2 calculations," *J. Chem. Phys.* **116**, 3175 (2002).

[32] Y. Jung, A. Sodt, P. M. W. Gill, and M. Head-Gordon, "Auxiliary basis expansions for large-scale electronic structure calculations," *Proc. Natl. Acad. Sci. U.S.A.* **102**, 6692 (2005).

[33] F. R. Manby, P. J. Knowles, and A. W. Lloyd, "The Poisson equation in density fitting for the Kohn-Sham Coulomb problem," *J. Chem. Phys.* **115**, 9144 (2001).

[34] M. Schütz and F. R. Manby, "Linear scaling local coupled cluster theory with density fitting I: 4-external integrals," *Phys. Chem. Chem. Phys.* **5**, 3349 (2003).

[35] T. Helgaker, P. Jorgensen, and J. Olsen, *Molecular electronic-structure theory*, Wiley, Chichester, 2000.

[36] P. Constans, P. Y. Ayala, and G. E. Scuseria, "Scaling reduction of the perturbative triples correction (T) to coupled cluster theory via Laplace transform formalism," *J. Chem. Phys.* **113**, 10451 (2000).

[37] M. E. Mura and P. J. Knowles, "Improved radial grids for quadrature in molecular density-functional calculations," *J. Chem. Phys.* **104**, 9848 (1996).

[38] C.-K. Skylaris, L. Gagliardi, N. C. Handy, A. G. Ioannou, S. Spencer, A. Willetts, and A. M. Simper, "An efficient method for calculating effective core potential integrals which involve projection operators," *Chem. Phys. Lett.* **296**, 445 (1998).

[39] T. D. Poulsen, K. V. Mikkelsen, J. G. Fripiat, D. Jacquemin, and B. Champagne, "MP2 correlation effects upon the electronic and vibrational properties of polyyne," *J. Chem. Phys.* **114**, 5917 (2001).

[40] S. Hirata and R. J. Bartlett, "Many-body Green's-function calculations on the electronic excited states of extended systems," *J. Chem. Phys.* **112**, 7339 (2000).

[41] M. Häser and J. Almlöf, "Laplace transform techniques in Møller–Plesset perturbation theory," *J. Chem. Phys.* **96**, 489 (1992).

[42] A. Takatsuka, S. Ten-no, and W. Hackbusch, "Minimax approximation for the decomposition of energy denominators in Laplace-transformed Møller–Plesset perturbation theories," *J. Chem. Phys.* **129**, 044112 (2008).

[43] D. Kats, D. Usvyat, S. Loibl, T. Merz, and M. Schütz, "Comment on "Minimax approximation for the decomposition of energy denominators in Laplace-transformed Møller–Plesset perturbation theories" [J. Chem. Phys. **129**, 044112 (2008)]," *J. Chem. Phys.* **130**, 127101 (2009).

[44] This name originates from the Fortran organization of multi-dimensional arrays in memory, for example $A(i, j)$ array elements are stored as follows $A(1, 1)$, $A(2, 1)$, ..., thus, index i is called the fastest index.

[45] S. Pissanetsky, *Sparse Matrix Technology*, Academic Press, London, 1984.

[46] G. W. Stewart, *Matrix Algorithms*, Society for Industrial and Applied Mathematics, Philadelphia, 1998.

[47] P. Y. Ayala and G. E. Scuseria, "Atom pair partitioning of the correlation energy," *Chem. Phys. Lett.* **322**, 213 (2000).

[48] GAUSSIAN Development Version, Revision E.02, M. J. Frisch, G. W. Trucks, H. B. Schlegel, G. E. Scuseria, *et al.*, Gaussian, Inc., Wallingford CT, 2004.

[49] K. N. Kudin and G. E. Scuseria, "A fast multipole algorithm for the efficient treatment of the Coulomb problem in electronic structure calculations of periodic systems with gaussian orbitals," *Chem. Phys. Lett.* **289**, 611 (1998).

[50] K. N. Kudin and G. E. Scuseria, "Linear-scaling density-functional theory with gaussian orbitals and periodic boundary conditions: Efficient evaluation of energy and forces via the fast multipole method," *Phys. Rev. B* **61**, 16440 (2000).

[51] J. P. Perdew, M. Ernzerhof, and K. Burke, "Rationale for mixing exact exchange with density functional approximations," *J. Chem. Phys.* **105**, 9982 (1996).

[52] R. Pino and G. E. Scuseria, "Importance of chain-chain interactions on the band gap of trans-polyacetylene as predicted by second-order perturbation theory," *J. Chem. Phys.* **121**, 8113 (2004).

[53] A. F. Izmaylov and G. E. Scuseria, "Efficient evaluation of analytic vibrational frequencies in Hartree–Fock and density functional theory for periodic non-conducting systems," *J. Chem. Phys.* **127**, 144106 (2007).

[54] S. Suhai, "Bond alternation in infinite polyene: Peierls distortion reduced by electron correlation," *Chem. Phys. Lett.* **96**, 619 (1983).

[55] M. Yu, S. Kalvoda, and M. Dolg, "An incremental approach for correlation contributions to the structural and cohesive properties of polymers. coupled-cluster study of trans-polyacetylene," *Chem. Phys.* **224**, 121 (1997).

[56] Please note that smaller basis sets (e.g. 3-21G) give rise to a substantial bond-length alternation with PBEh, but this result is obviously the consequence of the basis set incompleteness.

[57] D. Jacquemin, E. A. Perpète, G. Scalmani, M. J. Frisch, R. Kobayashi, and C. Adamo, "Assessment of the efficiency of long-range corrected functionals for some properties of large compounds," *J. Chem. Phys.* **126**, 144105 (2007).

[58] R. Pino and G. E. Scuseria, "Laplace-transformed diagonal Dyson correction to quasiparticle energies in periodic systems," *J. Chem. Phys.* **121**, 2553 (2004).

[59] M. Rohlfing, P. Krüger, and J. Pollmann, "Quasiparticle band-structure calculations for C, Si, Ge, GaAs, and SiC using gaussian-orbital basis sets," *Phys. Rev. B* **48**, 17791 (1993).

[60] M. Rohlfing and S. G. Louie, "Optical excitations in conjugated polymers," *Phys. Rev. Lett.* **82**, 1959 (1999).

[61] S. Grimme and F. Neese, "Double-hybrid density functional theory for excited electronic states of molecules," *J. Chem. Phys.* **127**, 154116 (2007).

[62] B. G. Janesko, T. M. Henderson, and G. E. Scuseria, "Long-range-corrected hybrids including random phase approximation correlation," *J. Chem. Phys.* **130**, 081105 (2009).

[63] C. R. Fincher, M. Ozaki, M. Tanaka, D. Peebles, L. Lauchlan, A. J. Heeger, and A. G. MacDiarmid, "Electronic structure of polyacetylene: Optical and infrared studies of undoped semiconducting $(CH)x$ and heavily doped metallic $(CH)x$," *Phys. Rev. B* **20**, 1589 (1979).

[64] L. Rossi, S. Alvarado, W. Riess, S. Schrader, D. Lidzey, and D. Bradley, "Influence of alkoxy substituents on the exciton binding energy of conjugated polymers," *Synth. Mat.* **111**, 527 (2000).

[65] A. Zunger, A. Katzir, and A. Halperin, "Optical properties of hexagonal boron nitride," *Phys. Rev. B* **13**, 5560 (1976).

chapter two

Density fitting for correlated calculations in periodic systems

Martin Schütz, Denis Usvyat, Marco Lorenz, Cesare Pisani, Lorenzo Maschio, Silvia Casassa, and Migen Halo

Contents

2.1   *Introduction*

In molecular computational chemistry, and in particular in post-Hartree–Fock (HF) applications, the density fitting (DF) approximation has proven to be one of the key techniques allowing large systems to be treated at a relatively high level of theory [3,6,11,12,15–17,33,35,43,48]. It permits a convenient factorization of the four-index electron repulsion integrals (ERIs), which makes the exploitation of disc space, memory, and CPU much more balanced and efficient and sometimes even reduces the scaling of the methods [36]. In periodic systems, the savings achieved by the introduction of the DF approximation are even more remarkable, due to the packing effects (especially in 3-D crystals) and the infiniteness of the studied objects.

Here we discuss precisely one such periodic application of DF. We have recently developed the CRYSCOR code [25, 29, 30, 41] which allows Møller–Plesset perturbation theory at second-order (MP2) calculations to be performed for crystalline systems, starting from the HF solution provided by the CRYSTAL code [9] in a basis set of Gaussian-type orbitals (GTO) centered on the nuclei (for other periodic MP2 approaches see e.g., [2,20,23,28,37,38] and Chapters 1, 3, and 4). The local-correlation approximation used here is an adaptation to the periodic case of the original Ansatz by Pulay and others [6, 32], which has been implemented in the MOLPRO package [19,34–36,48]. The introduction of DF in CRYSCOR has proven to be of primary importance for the efficient calculation of the ERIs between product distributions, since their exact evaluation through the four-index transformation of the integrals in the atomic orbital (AO) basis is prohibitively cumbersome in periodic systems.

In the following, we shall focus our attention on the evaluation of ERIs in CRYSCOR by means of DF techniques. In Section 2.2 we briefly review the formalism in the framework of LMP2 calculations for molecules, in order to introduce the problem and to specify notations. Two issues are treated in detail that are relevant in the present context: the massive use of Poisson-type orbitals (the Laplacian of GTOs), as fitting functions and the introduction of local fit-domains. Section 2.3 deals explicitly with the periodic problem, and discusses three DF approaches in this context. The first one is a straightforward extension of molecular DF with local fit-domains; the second one exploits the translational periodicity of the system, and reformulates DF in reciprocal space; the third approach is an appropriate combination of the other two. The performance of the three DF schemes is demonstrated in Section 2.4 for a few systems that cover a variety of typical situations: silicon and MgO, the prototype of pure covalent and pure ionic crystals, the ammonia molecular crystal, and two 2-D systems, the hexagonal BN monolayer, and a 3-layer MgO slab. This last system is used to explore the scaling of the computational cost with cell dimension, by considering supercells of increasing size. After assigning safe default values to some general parameters (the size of the fit-domains and the density of the k-net in the reciprocal fit), we explore the effect of the quality of the fitting set: for this purpose, we provide rules for defining VnZ sets, whose quality increase with n, by adapting standard molecular fitting sets to the periodic case. Finally, the results of a test computation for sodalite (a simple zeolitic structure) are reported to demonstrate the feasibility of LMP2 calculations for moderately complex periodic systems. Some general conclusions are tentatively drawn in Section 2.5.

2.2 DF in molecular LMP2 calculations

Consider the ERIs used in the LMP2 formalism:

$$K_{ia,jb} = (ia|jb) \equiv \int d\mathbf{r}_1 \int d\mathbf{r}_2 \frac{\rho_{ia}(\mathbf{r}_1)\rho_{jb}(\mathbf{r}_2)}{|\mathbf{r}_1 - \mathbf{r}_2|} \tag{2.1}$$

where $\rho_{ia}(\mathbf{r})$ and $\rho_{jb}(\mathbf{r})$ are density distributions, each being the product of an occupied localized molecular orbital (Wannier function—WF—in the periodic context) ϕ_i or ϕ_j and a projected atomic orbital (PAO) [5, 19, 32] ϕ_a or ϕ_b. Evaluation of these ERIs by the transformation from the AO basis forms the computational bottleneck of the conventional MP2 and LMP2 methods.

We now formulate the DF approximation by introducing a functional space $\{\chi\}$ with ket-vectors $|Q)$, each corresponding to a fitting function $\chi_Q(\mathbf{r})$, and the Coulomb metric $||\chi_Q||^2 = (Q|Q)$. Because of this choice of the metric, the bra vectors take the form:

$$(Q| = \int d\mathbf{r}' \frac{\chi_Q(\mathbf{r}')}{|\mathbf{r} - \mathbf{r}'|}. \tag{2.2}$$

If the basis vectors $|Q)$ are not mutually orthogonal in the Coulomb sense (as is usually the case), the projection operator onto the space of $\{\chi\}$ takes the form:

$$P^{\{\chi\}} = \sum_{Q'Q} |Q')[J^{\{\chi\}}]^{-1}_{Q'Q}(Q|, \tag{2.3}$$

with the metric matrix

$$J^{\{\chi\}}_{Q'Q} = (Q'|Q). \tag{2.4}$$

$\{\chi\}$ is referred to in the following as the auxiliary or fitting set, while the basis set of GTOs used for the linear combination of atomic orbitals (LCAO) expansion of molecular orbitals will be designated as the AO set. The factorization formula for the integral in Equation (2.1) is obtained by projecting, for example, the density ρ_{jb} onto $\{\chi\}$:

$$K_{ia,jb} \approx \tilde{K}_{ia,jb} = \sum_Q \left\{ \sum_{Q'} (ia|Q')[J^{\{\chi\}}]^{-1}_{Q'Q} \right\} (Q|jb)$$

$$= \sum_Q d^Q_{ia}(Q|jb). \tag{2.5}$$

In the case of a complete fitting set (or when the fitting basis entirely spans the space of the densities to be fitted, which is essentially a subspace of all possible pair-products of AOs), this approximation becomes an identity. In Equation (2.5) we introduced the notation of decorating the fitted quantities by a tilde.

A totally equivalent result is obtained when one looks at this technique from the angle of the fitting of the density distributions with an auxiliary set:

$$\rho_{ia}(\mathbf{r}) \approx \tilde{\rho}_{ia}(\mathbf{r}) = \sum_Q d_{ia}^Q \chi_Q(\mathbf{r}), \tag{2.6}$$

via the minimization of the error functional:

$$\Delta^{ia} = \int d\mathbf{r}_1 \int d\mathbf{r}_2 \frac{[\rho_{ia}(\mathbf{r}_1) - \tilde{\rho}_{ia}(\mathbf{r}_1)][\rho_{ia}(\mathbf{r}_2) - \tilde{\rho}_{ia}(\mathbf{r}_2)]}{|\mathbf{r}_1 - \mathbf{r}_2|}. \tag{2.7}$$

Commonly, GTOs centered on atoms are employed as fitting functions. Yet another fitting set, which we will essentially utilize in our method, allows a remarkable simplification of the 3- and 2-index integrals in Equation (2.5). It is usually called the Poisson-type orbital (PTO) set because of the close relation to the Poisson equation. Consider a set of GTOs $\{g_Q(\mathbf{r})\}$. The corresponding PTO set, $\{p_Q(\mathbf{r})\}$, is defined as follows [22, 26]:

$$p_Q(\mathbf{r}) = -\frac{1}{4\pi} \nabla^2 g_Q(\mathbf{r}). \tag{2.8}$$

Due to the identity (a form of the Poisson equation) [22]:

$$\int d\mathbf{r}' \frac{\nabla^2 g(\mathbf{r}')}{|\mathbf{r} - \mathbf{r}'|} = -4\pi g(\mathbf{r}) \tag{2.9}$$

the bra vectors (2.2) of type $p_Q(\mathbf{r})$ reduce to:

$$(Q| = \int d\mathbf{r}' \frac{p_Q(\mathbf{r})}{|\mathbf{r} - \mathbf{r}'|} = g_Q(\mathbf{r}). \tag{2.10}$$

This means that a Coulomb integral involving the function p_Q is essentially not a two-electron, but rather a one-electron integral involving the function g_Q instead. This implies not only a simplification of the integral evaluation, but also completely different decay properties of the integrals. The qualitative change in the decay rate becomes essential in periodic systems, where it is the key property that influences the convergence of the lattice sums. The PTOs possess an important property: they cannot hold a multipole moment of any order [21]. Thus, to use PTOs as fitting functions, one has to complement them with a few multipole-holding functions (e.g., GTOs), which would then fit the actual multipole moments of the densities [21].

In local-correlation methods, where the fitted densities are localized, a local density fitting (LDF) technique is often adopted, where instead of the full set of the fitting functions, only density-specific spatially restricted domains of the fitting functions (the so called *fit-domains*) are utilized. This technique is mandatory to achieve asymptotic linear scaling of the computational cost with the molecular size [48] in the DF applications. However, allowing the fit-domains $[ia]$ and $[jb]$ for the product distributions ρ_{ia} and ρ_{jb} from the integral $(ia|jb)$ to be different brings an additional, yet only technical, complication to the factorization formula for the 4-index ERI. As was suggested by Dunlap [10, 11] the error in the integral can be made second order with respect to the error in the density if one adopts the *robust* formalism of DF, i.e.:

$$\widetilde{K}_{ia,jb} = (\widetilde{ia}|jb) + (ia|\widetilde{jb}) - (\widetilde{ia}|\widetilde{jb}). \tag{2.11}$$

When the fitting set is the same for the bra and ket densities and the Coulomb metric has been employed, all three terms are identical, so that Equation (2.11) reduces to the usual one-term expression of Equation (2.5). However, in the LDF framework, all three terms of Equation (2.11) generally should be treated explicitly:

$$\widetilde{K}_{ia,jb} = \sum_{Q\in[ia]} d_{ia}^Q (Q|jb) + \sum_{Q\in[jb]} (ia|Q) d_{jb}^Q +$$

$$- \sum_{Q\in[ia]} \sum_{Q'\in[jb]} d_{ia}^Q (Q|Q') d_{jb}^{Q'}, \tag{2.12}$$

otherwise the robustness of the fitting cannot be guaranteed.

2.3 DF in periodic LMP2 calculations

The extension to periodic systems of the DF techniques outlined in the previous section is still rather scarce [14,25,39,40,45,46], and requires care. On the one hand, the infinite nature of the system prevents the straightforward (non-local) use of Equation (2.5), since the dimensions of the metric matrix to be inverted, $J^{\{x\}}$, would become infinite. On the other hand, translational symmetry, the essential property of crystals, opens new possibilities compared to the molecular case, such as the use of reciprocal-space techniques, the possibility of restricting the objects of the formalism to the reference cell, etc. Finally, the close packing of 3-D crystals poses new challenges concerning the size of the fit-domains and the treatment of the tails of localized quantities entering the formalism.

In the following, a *calligraphic* symbol will be used for identifying the cell to which each PAO, WF, fitting function, and so on, belongs. When a function is centered in the reference cell, the corresponding cell index is omitted. For example, $\rho_{ia\,\mathcal{A}}(\mathbf{r})$ corresponds to the product distribution $\phi_i(\mathbf{r})\,\phi_a(\mathbf{r} - \mathbf{R}_\mathcal{A})$, with the WF i in the zero cell, and the PAO a in the cell identified by the lattice vector $\mathbf{R}_\mathcal{A}$. After associating the label $\mathcal{B}_\mathcal{J}$ to the lattice vector $\mathbf{R}_\mathcal{B} + \mathbf{R}_\mathcal{J}$, we can also write: $\rho_{j\,\mathcal{J}b\mathcal{B}_\mathcal{J}}(\mathbf{r}) \equiv \rho_{jb\mathcal{B}}(\mathbf{r} - \mathbf{R}_\mathcal{J})$.

We describe below three different DF schemes that we have implemented in CRYSCOR and whose alternative use can be selected from input. The first one, *local direct-space* (LD), is an extension of molecular LDF [35,48] to the periodic case. This approach, similar to that adopted by Scuseria [14], is not expected to be very efficient for crystals with a small unit cell, due to the large fit-domains needed in 3-D-packed systems. At the same time, it exhibits the linear scaling features of the molecular formulation and thus becomes preferable when large unit cells are considered. The second scheme, *multipole-corrected-reciprocal* (MCR), fully exploits periodicity, since it is based on reciprocal space techniques. The problems related to the poor convergence of lattice sums, which affects the technique, can be partly circumvented by applying DF to the multipole corrected momentless distributions. Thirdly, a mixed scheme, *direct-reciprocal-decoupled* (DRD), is considered, which is formally more complicated, but embodies the most attractive features of the direct and reciprocal space techniques.

A common feature in these schemes is the full exploitation of the point group symmetry of the crystal, which permits all calculations to be restricted to the irreducible objects.

2.3.1 Local direct-space fitting in periodic systems

The LD technique results from the adaptation of the LDF molecular formalism (Equation [2.12]) to the periodic case. However, the strategy for the construction of the fit-domains deserves special attention, since their size has a severe impact on both accuracy and cost.

In the molecular LMP2 method with the LDF approximation [48], two variants of fit-domain definition have been proposed. In the former, a fit-domain $[ij]_\text{fit}$ is set to be the same for all the densities ρ_{ia} and ρ_{jb} corresponding to a given orbital [ij]-pair. In such a scheme, the fit-domain should contain all centers that support any of the possible ρ_{ia} and ρ_{jb} densities of this [ij]-pair with the PAOs from the corresponding pair-domain. Alternatively, fit-domains $[i]_\text{fit}$, being the union of all the $[ij]_\text{fit}$ fit-domains for a given i have been considered. In the density fitted Laplace transformed periodic MP2 method [2], which is applied to polymers, a single universal fit-domain for all the densities to be fitted is constructed [14].

All these approaches would imply in 3-D-packed periodic systems rather extensive fit-domains and thus inadequately expensive inversions of the fit-domain-specific metric matrices (see Equations [2.5] and [2.12]), which scale cubically with the number of the fitting functions. In order to circumvent this problem, we trade the size of the fit-domains for their number as in [35], where LDF is applied to the local coupled cluster with singles and doubles (LCCSD) method, but with a different procedure for the construction of the the fit-domains themselves. Namely, for each given pair of a WF and a PAO atomic center, a specific fit-domain is generated. This leads to much smaller fit-domains, cheap inversion, but a larger number of inversions to be done. This approach, as will be shown in the following, appears to be quite efficient and makes the pure LD technique competitive (rather unexpectedly) relative to the reciprocal space techniques even for small unit cells.

For accurate fitting, the fit-domains should be chosen such that the corresponding fitting functions provide sufficient support to the fitted densities. Therefore a simple and at the same time stable technique for estimating the population of the density to be fitted on a certain atom is needed. We define a quasi-population q_{DD}^{iaA} of the product density ρ_{iaA} on an atom D in cell $\mathcal{D}$ as:

$$q_{DD}^{iaA} = \sum_{\mu \in DD} [1 + P(i, a A)] \left(\sum_{\nu \in D\mathcal{D}} C_{\mu i} S_{\mu \nu} C_{\nu a A} \right)^2, \qquad (2.13)$$

where $C_{\mu a A}$ and $C_{\nu i}$ are the LCAO coefficients of PAOs and WFs, respectively, $\mathbf{S}$ is the AO overlap matrix and $P(i, a A)$ denotes the index permutation operator. Such a definition of the population slightly differs from the usual Mulliken form. The latter is not appropriate for our needs because the charge of any product density ρ_{ia} is zero due to the strict orthogonality between occupied and virtual spaces. Therefore an estimation of the support for such densities based on a standard population technique like Mulliken or Löwdin might become unstable in certain situations. In our approach, a fit-domain comprises a certain number N_D of atoms D with the highest populations q_{DD}^{iaA} for a block of PAOs belonging to a common center. As will be shown below, setting N_D to 10 atoms is usually sufficient.

2.3.2 Multipole-corrected-reciprocal fitting

Due to the reasons stated above (dense packing in 3-D bulk crystals, the need for large fit-domains, thus leading to costly matrix inversions) it may appear convenient to reformulate DF in reciprocal space.

Let us define a discrete Monkhorst mesh, sampling the Brillouin zone in N_k points, and define the Fourier image of the fitted ERIs,

$$\widetilde{K}_{iaA, jbB}(\mathbf{k}) = \sum_{QQ'} (i\, a\, \mathcal{A} | Q)_{\mathbf{k}} [J^{\{x\}}(\mathbf{k})]_{QQ'}^{-1} (j\, b\, \mathcal{B} | Q')_{\mathbf{k}}^* \qquad (2.14)$$

by means of Fourier transform (FT)/back Fourier transform (BFT) of the involved quantities [25, 30]:

$$J_{QQ'}^{\{x\}}(\mathbf{k}) = \sum_{Q'} (Q|Q'Q') \exp(i\mathbf{k}\mathbf{R}_{Q'}),\tag{2.15}$$

$$(i\,a\,\mathcal{A}|Q)_{\mathbf{k}} = \sum_{Q} (i\,a\,\mathcal{A}|QQ) \exp(i\mathbf{k}\mathbf{R}_{Q}),\tag{2.16}$$

and, finally,

$$\widetilde{K}_{i\,a\,\mathcal{A},\,j\,\mathcal{J}\,b(\mathcal{B}_{\mathcal{J}})} = \frac{1}{N_{\mathbf{k}}} \sum_{\mathbf{k}} \widetilde{K}_{i\,a\,\mathcal{A},\,j\,b\mathcal{B}}(\mathbf{k}) \exp(-i\mathbf{k}\mathbf{R}_{\mathcal{J}}),$$

$$\text{with } \mathcal{B}_{\mathcal{J}} \text{ defined as } \mathbf{R}_{\mathcal{B}} + \mathbf{R}_{\mathcal{J}}.\tag{2.17}$$

This approach allows all matrix multiplications and inversions to be reduced to the size of the fitting set in the crystalline unit cell (in k-space the fitting function as well as both WF i- and j-indices are restricted to the reference cell; see [25] for discussion). However, it exhibits the following difficulties: (i) the FT of the ERIs (Equation [2.15]) and Equation [2.16] imply slowly converging infinite lattice sums; (ii) since the resulting 4-index ERIs are needed in the direct space, their slow decay demands a large number of k-points in the BFT (Equation [2.17]) and thus in the reciprocal fitting. These issues can be nicely overcome by the use of PTOs, which, as we already mentioned in Section 2.2, do not hold any multipole moment of any order and thus Coulomb integrals involving them fade exponentially with distance.

Since the PTOs alone can fit only momentless densities, we define for each product distribution $\rho_{ia\,\mathcal{A}}$ a *correcting distribution* $\Pi_{ia\,\mathcal{A}}$, which is a linear combination of a *correcting set* of GTOs, situated in a local domain $[ia\,\mathcal{A}]$ of a few atoms around the center of $\rho_{ia\,\mathcal{A}}$:

$$|\Pi_{ia\,\mathcal{A}}) = \sum_{G\in[ia\,\mathcal{A}]} c_{ia\,\mathcal{A}}^{G}|G).\tag{2.18}$$

This distribution is then subtracted from the corresponding ρ_{ia} distribution, in order to obtain a corrected, momentless (up to a certain order) distribution,

$$\rho'_{ia\,\mathcal{A}} = \rho_{ia\,\mathcal{A}} - \Pi_{ia\,\mathcal{A}},\tag{2.19}$$

which then can conveniently be fitted by PTOs only. The fitting is thus performed in reciprocal space, according to Equations (2.14)–(2.16), for the corrected density distribution, leading after BFT (Equation [2.17]) to the momentless fitted integrals $\widetilde{K}'_{ia\,\mathcal{A},\,j\mathcal{J}b\mathcal{B}_{\mathcal{J}}}$ in the direct space. The final fitted integrals are obtained by adding back the contribution of the correcting distribution to the integral:

$$\widetilde{K}_{ia\,A,\,jJbB_J} = \widetilde{K}'_{ia\,A,\,jJbB_J} + (\rho_{ia\,A}|\Pi_{jJbB_J}) +$$
$$+ (\Pi_{ia\,A}|\rho_{jJbB_J}) - (\Pi_{ia\,A}|\Pi_{jJbB_J}). \tag{2.20}$$

We still need to define how the c_{ia}^G of Equation (2.18) are to be determined. The purpose of the $\Pi_{ia\,A}$ function is to smoothly represent a part of the corresponding distribution holding its multipole moments, in order to simplify the consecutive PTO fit. This can be accomplished by performing a fitting procedure within the small correcting GTO basis, combined with additional constraints for the multipoles. The resulting functional is analogous to that of Equation (2.7), but with additional terms making sure that all multipoles up to a given order must be the same for $\Pi_{ia\,A}$ and $\rho_{ia\,A}$:

$$\Delta^{\Pi_{ia\,A}} = \int d\mathbf{r}_1 \int d\mathbf{r}_2 \frac{[\rho_{ia\,A}(\mathbf{r}_1) - \Pi_{ia\,A}(\mathbf{r}_1)][\rho_{ia\,A}(\mathbf{r}_2) - \Pi_{ia\,A}(\mathbf{r}_2)]}{|\mathbf{r}_1 - \mathbf{r}_2|} -$$
$$+ \sum_l \lambda_{ia\,A}^l \left[\left(\sum_{G \in [ia\,A]} c_{ia\,A}^G m_G^l \right) - m_{ia\,A}^l \right], \tag{2.21}$$

where l indicates the multipole moment ranging from 0, i.e., charge, up to l_{max}. The $\lambda_{ia\,A}^l$ are Lagrange multipliers and the $m_{ia\,A}^l$, m_G^l are the true multipole moments of the $\rho_{ia\,A}$ distribution and the individual GTOs, respectively. The linear equation system that minimizes this functional can schematically be displayed as follows:

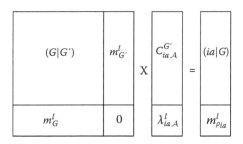

The evaluation of the coefficients practically implies an inversion of a symmetric matrix, in the same way as in the standard local fit.

The MCR technique just described is a generalization of the dipole-correction approach we initially developed for the periodic DF in [25], and is in principle the fastest among those proposed here. However, it exhibits some drawbacks, which have their origin in the fact, that this kind of fitting is not robust according to Dunlap's definition (Equation [2.11]). Indeed, the error in the integral is second order with respect to the error $\delta'_{ia} = \rho'_{ia} - \tilde{\rho}'_{ia}$ in the corrected (momentless) density, but not with respect to the error $\delta_{ia} = \rho_{ia} - \tilde{\rho}_{ia}$ in the true density:

$$
\begin{aligned}
K_{ia,jb} &= (\rho'_{ia}|\rho'_{jb}) + (\rho_{ia}|\Pi_{jb}) + (\Pi_{ia}|\rho_{jb}) - (\Pi_{ia}|\Pi_{jb}) = \\
&= \left[\tilde{K}'_{ia,jb} + (\delta'_{ia}|\delta'_{jb})\right] + (\rho_{ia}|\Pi_{jb}) + (\Pi_{ia}|\rho_{jb}) - (\Pi_{ia}|\Pi_{jb}) \\
&= \tilde{K}_{ia,jb} + (\delta'_{ia}|\delta'_{jb}),
\end{aligned} \tag{2.22}
$$

where the lattice vector index has been dropped for simplicity. In the derivation we utilized Equations (2.19) twice, as well as (2.20) and the robustness of the PTO-fit for the corrected integral $(\rho'_{ia}|\rho'_{jb})$.

To show how this non-robustness of the fit can lead to instabilities, let us analyze in detail what happens when the ERIs, which are computed in the redundant PAO basis (a, b), are transformed to the local orthogonal basis $(\bar{a}, \bar{b})$, specific for a given pair [ij], which takes place when the LMP2 equations are solved [30]. The transformation:

$$
\mathbf{U}\{K_{ia,jb}\} = \{K_{i\bar{a},j\bar{b}}\} \tag{2.23}
$$

is applied, where $\mathbf{U}$ is constructed from the eigenvectors of the PAO pair-domain specific overlap matrix. The eigenvalues of this matrix represent the norm of the transformed orthogonal functions. Zero or close-to-zero eigenvalues indicate redundant functions. Therefore the eigenvectors corresponding to such eigenvalues are excluded from the transformation matrix $\mathbf{U}$. The threshold for the largest norm of the functions to be considered redundant should neither be set too low, in order to maintain numerical stability, nor too high, in order not to spoil the quality of the virtual space. In our calculations we usually employ a value of $\bar{T} = 10^{-4}$ for this threshold.

When the transformation of Equation (2.23) is applied to the fitted ERIs we have:

$$
\begin{aligned}
\mathbf{U}\{\tilde{K}_{ia,jb}\} &= \mathbf{U}\{K_{ia,jb} - (\delta'_{ia}|\delta'_{jb})\} = \\
&= \{K_{i\bar{a},j\bar{b}} - (\delta'_{i\bar{a}}|\delta'_{j\bar{b}})\}.
\end{aligned} \tag{2.24}
$$

The exact integrals $K_{i\bar{a},j\bar{b}}$, which correspond to the virtuals with a small norm, become quite small themselves. However, there is no guarantee that the term $(\delta'_{i\bar{a}}|\delta'_{j\bar{b}})$, and thus the fitted integrals $\tilde{K}_{i\bar{a},j\bar{b}}$, which are actually used in the DF calculations, are also small. In fact, the error distribution $\delta'_{i\bar{a}}$ refers to the corrected density $\rho'_{i\bar{a}}$. The latter is formed according to the transformation $\mathbf{U}$ from the correcting functions Π_{ia}, which fit the densities ρ_{ia} rather roughly. As a result, the superposition $\Pi_{i\bar{a}}$ of the corrections Π_{ia} and thus the corrected density $\rho'_{i\bar{a}}$ can be significantly "larger" than the actual density $\rho_{i\bar{a}}$. The error term $(\delta'_{i\bar{a}}|\delta'_{j\bar{b}})$ is small with respect to the absolute value of the integral $K'_{i\bar{a},j\bar{b}}$, but, as follows from Equation (2.24) and the above analysis, not necessarily so compared to $K_{i\bar{a},j\bar{b}}$. Finally, this error is blown up further by the subsequent normalization of the functions $(\bar{a}, \bar{b})$.

This instability manifests itself in the relatively poor performance of the fitting basis sets of moderate sizes and in a strong dependence of the results on the choice of the local fitting domains and can be reduced at the price of more expensive calculations. Indeed, in order to make the corrected density similar to the actual density in the local orthogonal basis, the correcting moment-constrained fit should be as smooth as possible. This can be achieved by improving the correcting basis set, using e.g., some of the PTOs for the correction rather than the reciprocal fit, and increasing the size of the fit-domains. In addition, the quality of the reciprocal fit basis should also be rather high, in order to reduce the effect of non-robustness of the fit. This issue is investigated further numerically in Section 2.4.

2.3.3   Direct-reciprocal-decoupled fitting

Recently we proposed a periodic DF method that does not require a multipole correction or constraining (although the latter can optionally be included), and at the same time provides robust fitting within the fitting basis set chosen [24]. In this method, the basis set is divided into two parts: a rich PTO basis spanning the whole crystal, labeled as $\mathbf{P}$, and a density-specific small GTO basis, to fit the multipole moments, labeled as $\mathbf{G}$. However, a straightforward application of this scheme is not possible, since the metric matrix cannot be inverted in either reciprocal or direct space (in the former case, one would need to transform the slowly decaying GTO-based integrals into the reciprocal space, while for the latter, one would have to deal with a metric matrix of infinite dimensions due to the extended PTO-part of the basis). In order to decouple the two sets, we project the GTO part of the auxiliary basis onto the space complementary to that spanned by the PTOs, using the Coulomb-projection operator (Equation [2.3])

$$|\Gamma) = |G) - \sum_{PP'} |P)[J^{\{P\}}]^{-1}_{P,P'}(P'|G). \tag{2.25}$$

The projected GTO's Γ can now be used as the direct space part of the fitting basis instead of the GTO's $\mathbf{G}$. The block diagonal form of the metric matrix **J**, with zero off-diagonal blocks **Γ-P**:

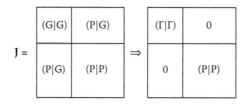

now permits an independent inversion of both diagonal blocks: the PTO-block in the reciprocal space, and **Γ**-block in the direct space. This technique is denoted as the *direct-reciprocal-decoupled* (DRD) DF.

The functions **Γ** serve as moment-fitting functions completely in the same manner as GTOs, since they differ from the latter only by a linear combination of momentless PTOs. Moreover, the quality of the fit does not deteriorate when going from the GTOs to the functions **Γ**, since the projection of Equation (2.3) does not reduce or modify the space spanned by the initial fitting basis. The **Γ**-fit is performed in the same way as the direct-space fit of Section 2.3.1, but with the projected fitting functions **Γ** instead of the usual fitting functions and with much smaller fit-domains. Large fit-domains are in fact not needed, since the fitting basis has a relatively rich reciprocal part, which is not restricted to any local domain.

Once the fit-domains are defined, one has to evaluate the integrals $(\Gamma|\Gamma)$ and $(ia|\Gamma)$. Using the definition of the functions **Γ** (Equation [2.25]) yields:

$$(\Gamma|\Gamma) = (G|G) - \sum_{PP'}(G|P)[J^{\{P\}}]^{-1}_{P,P'}(P'|G) \qquad (2.26)$$

and

$$(ia|\Gamma) = (ia|G) - \sum_{PP'}(ia|P)[J^{\{P\}}]^{-1}_{P,P'}(P'|G). \qquad (2.27)$$

The first terms in the right-hand side (RHS) of Equations (2.26) and (2.27) are the usual integrals with the GTOs, while the second terms can be interpreted as the equivalent integrals, but fitted by the PTO-only fitting basis. As discussed above, this fit can be carried out only in the reciprocal space. Therefore, one has to employ the very same machinery of the PTO reciprocal-space fit as for the actual 4-index ERIs using Equations (2.15), (2.16), and (2.17).

The DRD scheme, when applied to very closely packed systems and with rich PTO fitting basis sets (say of quadruple zeta quality), might exhibit slight numerical instabilities which manifest themselves in slow k-mesh convergence and in the need for very tight thresholds in the 3-index integral evaluation. We have recently investigated this problem [42] and found that it can be circumvented by including the most diffuse part of the PTO basis in the direct space together with the GTOs.

2.4 Test calculations

2.4.1 Fitting basis sets

PTO sets optimized for different AO sets in the molecular context have been reported [4, 31]. However, the number of such basis sets is rather limited and besides, in periodic systems, an optimized AO set for a given atom can significantly differ from one compound to another depending on the structure and the type of binding. In order to avoid the issue of re-optimizing the fitting basis sets for each particular study, we adopt the following strategy. We take the exponents and contraction coefficients of existing molecular GTO fitting sets, optimized for an AO set of a certain quality (e.g., those of the cc-pVXZ family [47]), and assign them to the PTO part of the auxiliary basis. We use a common set of fitting GTOs, one shell per each angular momentum and each center, with the following exponents: 0.3, 0.5, 0.7, and 0.9 Bohr^{-2} for p-, d-, f-, and g-type GTOs, respectively. Note that s-type GTOs are not needed in principle since the densities to be fitted in the LMP2 method are chargeless; however, their inclusion (with a 0.3 exponent) in the MCR approach, contributes to the performance of the fitting set in this case. The resulting fitting sets are indicated below as VDZ, VTZ, VQZ, V5Z, for analogy with the molecular notation.

Since the CRYSTAL code, which provides the HF reference, is not able to treat AOs beyond the f-type in the orbital basis, we do not go beyond g-functions in the fitting set.

2.4.2 General computational parameters

The selected test systems are presented in Table 2.1; the geometries are in all cases the experimental ones. AO basis sets of triple-zeta quality of type [3s2p1d] for H, [4s3p2d1f] for B,N,O, and Mg, and [5s4p2d1f] for Si atoms are employed. Since the periodic LCAO self-consistent field (SCF) calculations may suffer from quasi-linear dependencies even for basis sets of triple-zeta quality, the standard molecular basis sets are not always applicable as they are. For H and first-row elements, Pople's 6-311G(2d,1f) sets [18] have been used with slight modifications to the Boron basis set (the exponents of the two most diffuse s and p functions are raised from 0.315454, 0.0988563 to 0.60, 0.25 Bohr^{-2}, while that of the external d function from the original 0.2005 to 0.25 Bohr^{-2}). For Mg and Si, the s- and p-orbitals have been taken from Refs. [44] and [27], respectively, complemented by d- and f-orbitals from Pople's (2d,1f)-sets (with f-orbital for Si upscaled to 0.45 Bohr^{-2}). All our MP2 calculations employ the frozen core approximation.

Table 2.1 Main Characteristics of Test Systems

System	Group	a	N_{at}	N_{WF}	Domain
Si	$Fd\bar{3}m$	5.43	2	4	$Si_3Si\text{-}SiSi_3$
MgO	$Fm\bar{3}m$	4.21	2	4	$O[Mg_6]$
NH_3	$P2_13$	5.048	16	16	NH_3
BN-m	$P\bar{6}m2$	2.504	2	4	$N[B_3][N_6]$
$(MgO\text{-}3m)_n$	$P4/mmm$	2.977	$6n$	$12n$	$O[Mg_x]$

All systems are 3-D cubic crystals except for BN-m, which is a monolayer of hexagonal symmetry, and MgO-3m, a 3-layer slab of tetragonal symmetry, cut out from the bulk: in the last case, along with the reference case (S_1) with unit vectors a_i, two supercells have been considered, S_2 and S_3, with unit vectors $(a_1 \pm a_2)\sqrt{2}$ and $2a_i$, respectively. Lattice parameters a are in Å, and correspond to experimental equilibrium geometries; the N-H bond length in NH_3 is 1.0219 Å [13]. N_{at} is the number of atoms per cell. In the last two columns, the number of WFs per cell (half the number of valence electrons) and the composition of their excitation domains (type and number of atoms) are reported: in the MgO-3m slab, x is 6 for WFs centered on oxygens in the central layer, and 5 otherwise.

The WF domains (that is, the set of atoms individualizing the PAOs to which electron excitations are allowed from a given WF pair) are listed in Table 2.1. They include up to first or second nearest neighbors about the WF center, except for the ammonia crystal where the domains coincide with the molecular units. The *core domains* are the Boughton-Pulay domains [5] (defined with a threshold of 0.98) and usually comprise one atom for atom-centered WFs and two atoms for bond-centered WFs. The tails of WFs and PAOs are truncated by disregarding the support from all centers whose AO coefficients drop below a threshold $c_{tail} = 0.0001$. Redundancies of the PAOs are treated by excluding the eigenvectors of the pair-specific PAO overlap matrices, whose eigenvalues are below a threshold of $\overline{T} = 0.0001$ (see Section 2.3.2). Finally, the cut-off interorbital distance (the distance between the core domains of every pair of the WFs) for ERIs calculated by using the DF approximation was set to 6 Å. Since we focus rather on the DF aspects of the LMP2 calculations, we do not consider pairs beyond 6 Å. The exceptions are the MgO slab and the sodalite, where we include pairs from 6 Å to 11 Å by exploiting the multipole expansion and (only in sodalite) estimate the effect of the remaining pairs up to infinite distance by the C_6R^{-6} extrapolation [30]. In the DRD and MCR schemes, for the reasons mentioned above, all PTOs with exponents less than 0.9 $Bohr^{-2}$ are treated together with GTO-fitting functions in the direct space.

All of the timings reported in the following correspond to non-dedicated non-parallel calculations performed on a single core of an 8-processor Xeon node with a clock rate of 3000 MHz.

2.4.3 DF accuracy criteria

The primary criterion for judging the accuracy of the approximate evaluation of the ERIs within the selected range is simply the convergence of the LMP2 energy with increasing quality of the fitting (size and quality of fitting set, size of fitting domains, etc.).

A more stringent criterion is based on a comparison between the exact ERIs and their DF estimate. However, the complete exact calculation is impossible in practice; for instance, the evaluation of the exact ERIs for a *single* WF pair in BN-m takes 95 hours, to be compared to several minutes for the corresponding DF calculation.

We therefore restrict this comparison in the following to a few subsets of ERIs in the case of BN-m. A label z individualizes the general subset, $\mathcal{E}_z$, characterized by four indices: $i(z)$, $j(z)$ (a given WF pair), and $A(z)$, $B(z)$ (a pair of atoms in the $[ij]$ domain). The corresponding $\mathcal{E}_z$ set comprises all $(ia|jb)$ ERIs where $i = i(z)$, $j = j(z)$ and a, b are PAOs belonging to atoms $A(z)$ and $B(z)$, respectively. In the present case, the number N_z of integrals in each $\mathcal{E}_z$ set is about 500.

The quality of the fit is estimated in terms of the mean absolute deviation $\overline{\Delta}_z = \sum_{ia,jb \in \mathcal{E}_z} |\widetilde{K}_{ia,jb} - K_{ia,jb}|/N_z$ of the fitted integrals with respect to the exact ones in the given subset. The average absolute value of the ERIs in the selected subset: $\overline{R}_z = \sum_{ia,jb \in \mathcal{E}_z} |K_{ia,jb}|/N_z$, allows us to estimate the relative error.

2.4.4 Adjustment of DF parameters

The three DF approaches described above though methodologically quite different, possess some common features. For all of them, direct-space fit domains (though of different sizes) have to be introduced. The last two, involving the reciprocal space fit, also need the k-mesh to be defined. We explore here the influence of these settings on the accuracy of the fit in order to find optimal values for them.

Table 2.2 reports results obtained using the LD scheme with the first four test systems. The minimal number of atoms in the fit-domains N_D (which is automatically increased by the program, if this number does not preserve the point-symmetry invariance of the fit-domains) is varied over a large range. Evidently, convergence is rapidly reached in all cases. In the following, we use an N_D value of 10 atoms in the LD method. For the DRD scheme, the fit-domain sizes are of less importance, since a large part of the fitting basis is processed in the reciprocal space and thus does not have any spatial restriction. In our calculations we will use the value $N_D = 6$, which is usually sufficient. The stability of the DRD and MCR schemes in this respect is compared in Section 2.4.5.

Table 2.2 Dependence of the LMP2 Energy (Hartree) from the Parameter N_D, the Minimal Number of Atoms Included in Each Local Fit Domain

N_D	Si	MgO	NH_3	BN-m
5	−0.233525	−0.282436	−0.983313	−0.277011
10	−0.233516	−0.282323	−0.983321	−0.276838
15	−0.233517	−0.282323	−0.983329	−0.276839
20	−0.233513	−0.282327	−0.983326	−0.276840
30	−0.233519	−0.282328	−0.983333	−0.276840

All calculations were done with the LD technique, using a VTZ fitting set, except for NH_3, where a VQZ set was used.

The dependence of the results on the density of the k-mesh employed in the reciprocal part of the DRD fitting (Equations [2.14]–[2.17]) is shown in Table 2.3. Too loose nets result in catastrophic behavior. However, convergence is rapidly reached: for systems with small unit cells (MgO and Si crystals, BN slab) a shrinking factor of 6, corresponding to 216 and 36 k points for 3- and 2-D systems, respectively, is totally adequate and is used as the default value in the following. Moreover, the larger the unit cells, the more sparse k-meshes can be utilized, such that for the ammonia crystal with 16 atoms per unit cell, only a $3 \times 3 \times 3$ k-mesh is needed.

2.4.5   *Performance of the three DF schemes*

As anticipated in Section 2.3.2, the MCR scheme is more sensitive to computational settings than the other two. We demonstrate this undesirable feature by analyzing its performance in the case of silicon.

Table 2.3 Dependence of the LMP2 Energy (Hartree) from the Shrinking Factor s_k^{fit}, which Determines the Density of the k-mesh in the Reciprocal Part of DRD Fitting

s_k^{fit}	Si	MgO	NH_3	BN-m
2	—	—	−0.983479	—
3	—	—	−0.983317	—
4	−0.243413	−0.282324	−0.983317	—
5	−0.233520	−0.282314	−0.983317	−0.278602
6	−0.233519	−0.282314	−0.983317	−0.276838
7	−0.233519	−0.282314	−0.983317	−0.276838

All these DRD calculations were done using a VTZ fitting set, except for NH_3 where a VQZ set was used, and with the minimum number of atoms in the direct part set to 6.

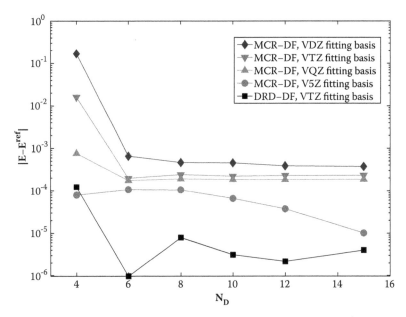

Figure 2.1 Difference, in a logarithmic scale, between the LD-fitted LMP2 energy taken as the reference (see text), and that calculated with MCR using different fitting sets, as a function of the number of atoms in the fit-domains, N_D. For comparison, the corresponding data for a DRD calculation with a VTZ fitting set are reported.

In Figure 2.1 the dependence of the accuracy of the MCR and DRD fits is shown using fit-domains of different sizes. Since the exact calculation of the ERIs is not feasible for this system and basis set, the energy provided by the LD fitting scheme with N_D set to 16 atoms in the fit-domains and the V5Z fitting set has been taken as the reference. This can be considered as a reasonably converged value since, as is seen from Figure 2.1 and confirmed from the other data reported below, all three DF methods tend to the same result within a few μHartree discrepancy, which is far below the desired accuracy of the DF approximation itself. Figure 2.1 clearly demonstrates the problems of the non-robust MCR fitting. The convergence with respect to the fitting basis set is very slow, and only the V5Z curve finally approaches the correct value, while the error produced when using the lower-quality fitting sets is still above 10^{-4}. This pattern was also observed in our previous calculations with the dipole correction scheme [41] where we had to choose quite rich fitting basis sets and a loose $\overline{T}$ threshold in order to achieve reasonable accuracy. Moreover, the convergence of the MCR-DF results with the fit-domain size is rather poor, as expected: the V5Z result provides the desired accuracy only with fit-domains consisting of 10 atoms or more. The same behavior is observed with other fitting sets,

Table 2.4 The LMP2 Energies Per Cell (Hartree) from Pairs in the Selected Range Reported Along with the Elapsed Times in Seconds (in Parentheses)

System	Si	MgO	NH$_3$	BN-m
Fitting set		Local direct-space fitting		
VDZ	−0.234288	−0.284063	−0.984367	−0.277720
	(3889)	(2505)	(6117)	(786)
VTZ	−0.233516	−0.282323	−0.983145	−0.276838
	(7587)	(3090)	(9984)	(953)
VQZ	−0.233514	−0.282316	−0.983321	−0.276843
	(7728)	(4276)	(15663)	(1596)
V5Z	−0.233518	−0.282321	−0.983347	−0.276848
	(8774)	(5393)	(21870)	(2076)
		Direct-reciprocal decoupled fitting		
VDZ	−0.234335	−0.284183	−0.984439	−0.277811
	(3623)	(2508)	(5268)	(843)
VTZ	−0.233519	−0.2823231	−0.983141	−0.276844
	(5762)	(2816)	(8975)	(958)
VQZ	−0.233513	−0.282315	−0.983317	−0.276847
	(6354)	(3629)	(13663)	(1017)
V5Z	−0.233519	−0.282322	−0.983366	−0.276851
	(7795)	(3739)	(15124)	(1119)

Fit-domains of 10 and 6 atoms have been employed in the LD and DRD schemes, respectively. In the DRD method $6 \times 6 \times 6$ k-mesh has been utilized for the Si and MgO crystals, $3 \times 3 \times 3$ for the NH$_3$ crystal, and 6×6 for the BN slab.

but the fluctuations are "hidden" in the logarithmic scale of Figure 2.1. A much faster convergence with the fit-domains' size is seen for the DRD fit. The error now lies in the μHartree region, which we consider as rather fortuitous or connected to a possible deficiency in the reference value. More importantly, the convergence of energy within a few μHartrees is already reached at $N_D = 6$, that is, much faster than in the case of the MCR fit. These results clearly indicate that DRD and LD fits are superior compared to the MCR scheme, and that the robustness of the fit is essential in our approach.

Table 2.4 compiles LMP2 energies and timings for the first four test systems obtained with the LD and DRD schemes, using fitting sets of progressively better quality. All other parameters are set to their default values. Notice that, apart from the VDZ case, all other fitting sets perform very well, providing results coinciding within a few μHartree/atom. The scaling of the timings with the size of the fitting set is quite reasonable: a factor of about 1.5 is observed when going from VTZ to V5Z. The reason for this

Table 2.5 Average Absolute Value ($\bar{R}_z$) and Mean Absolute Deviation of the Fitted set from the Exact Value ($\bar{\Delta}_z$) of the ERIs in Six Different $\mathcal{E}_z$ Subsets, as a Function of the Fitting Set

z	$i(z)$	$A(z)$	$j(z)$	$B(z)$	$\bar{R}_z$	$\bar{\Delta}_z^{VDZ}$	$\bar{\Delta}_z^{VTZ}$	$\bar{\Delta}_z^{VQZ}$	$\bar{\Delta}_z^{V5Z}$
1	w_1^1	$N_{(1)}$	w_1^1	$N_{(1)}$	14369040	131171	7292	3445	2323
						128573	7852	3476	2320
2	w_1^1	$N_{(1)}$	w_1^1	$B_{(1)}$	3867580	17321	1317	920	820
						17996	1361	927	819
3	w_1^1	$N_{(1)}$	w_2^1	$B_{(1)}$	2233300	22713	1048	921	857
						23940	1122	928	870
4	w_1^1	$N_{(1)}$	w_3^2	$B_{(2)}$	105801	1683	352	325	323
						1041	331	322	321
5	w_1^1	$N_{(1)}$	w_3^3	$N_{(3)}$	73669	1200	297	284	283
						748	294	284	283
6	w_1^2	$N_{(1)}$	w_3^z	$N_{(3)}$	75636	689	59	43	39
						702	59	43	40

All data is in nanoHartree. The subsets are identified by means of the $i(z)$, $A(z)$, $j(z)$, and $B(z)$ indices as explained in the text. The first line for each z set refers to LD, the second to DRD calculations.

small price is attributed to the fact that basically the PTO part is expanded, while the bottleneck of the calculations presently lies in the evaluation of the 3-index two-electron integrals over the GTO part of the fitting basis. Finally, the two DF techniques appear as almost equivalent, concerning both results and efficiency, with the DRD fit being slightly faster when large fitting basis sets are employed.

Consider now the second criterion for evaluating the accuracy of the DF approximation introduced in Section 2.4.3, based on the comparison between exact and estimated ERIs. The results reported in Table 2.5 concern six subsets $\mathcal{E}_z \equiv \{i(z)A(z)j(z)B(z)\}$ of the BN-m system. The notation adopted for identifying each subset is as follows, with reference to the scheme below:

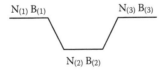

In the Ith cell, the four valence WFs are all centered on the $N_{(I)}$ nucleus; one of them (w_1^2) has the same symmetry as the p_z orbital on that atom, the other three (w_1^1) describe symmetry-equivalent bond-type orbitals directed toward one of the neighboring $B_{(J)}$ atoms. For instance, the third subset involves a pair of WFs (w_1^1, w_2^1) centered on neighboring nitrogen atoms ($N_{(1)}$, $N_{(2)}$), and directed toward the boron atom $B_{(1)}$ in between; the PAOs of the first product distribution belong to $N_{(1)}$, those of the second to $B_{(1)}$.

The values of $\overline{R}_z$ (the average absolute value of the integral) and of $\overline{\Delta}_z^{VnZ}$ (the average absolute value of the discrepancy for the n-th fitting set) are reported in Table 2.5 in nanoHartree. $\overline{R}_z$ generally decreases along the z series since the subsets correspond to progressively more distant product distributions, and thus to smaller ERIs. So, for instance, the two WFs in the first two subsets are located in the reference cell, in the third they are at a distance of 1.58 Å, in the fourth at 2.50 Å, and in the last two at 4.34 Å, respectively. The dependence of $\overline{\Delta}_z$ on the fitting basis is given in Table 2.5 for both the LD and the DRD method. It is seen that, for all z's and both methods, there is an impressive improvement when passing from the VDZ to the VTZ fitting set, whereas along the VTZ-VQZ-V5Z series the improvement is appreciable only for the first two integral sets involving "strong" WF pairs (zero distance between the WFs). This implies that the choice of good-quality fitting sets mainly affects strong pairs (actually those that contribute the most to the final energy), and only marginally affects the others. The agreement factor, i.e., the ratio between $\overline{\Delta}_z$ and $\overline{R}_z$ is excellent—a few parts in 10^4 for all fitting sets except VDZ; it is practically constant along the z series and about the same for the two different DF schemes.

Table 2.6 demonstrates the stability of the solution (in the sense of the size extensivity of the method) and provides an analysis of the scaling of the computational time with the supercell size in the case of the MgO-3m slab. This study can also be considered as prototypical for modeling surface adsorption, where slabs with unit cells of increasing sizes are often studied. The k-meshes in the DRD-DF calculations were adjusted to the supercell size, i.e., the 6×6 for the initial cell case, then 4×4 for the doubled supercell and 3×3 for the quadrupled one.

First of all we note that the HF reference is stable and the computational cost scales linearly, which is rather unusual for the HF method. In order to provide an accurate Fock matrix we used very tight thresholds for pre-screening the exchange integrals: 20 for both ITOL4 and ITOL5 [9]. Therefore the main fraction of the HF computational time was spent in the evaluation of the exchange part of the Fock matrix, the computational cost of which indeed scales linearly with respect to the number of atoms in the cell.

The localization and symmetrization of the Wannier functions demonstrates a close-to-cubic scaling, and in combination with a non-optimal performance, can become a bottleneck for systems with a large unit cell. Significant effort is presently being invested to remedy this issue [49].

The LMP2 part demonstrates very good performance, and most importantly, a nearly linear scaling of the computational cost. The computational

Table 2.6 Stability of the Solution and Scaling of the Computational Times with Respect to the Supercell Size for MgO-3m Slabs

		6 Atoms	12 Atoms	24 Atoms
Elap. time	Hartree–Fock	3210	6748	12666
	Localization/ Symmetrization	175	1286	11673
	E_{HF}	−823.991859	−823.991861	−823.991862
	Direct-reciprocal decoupled DF			
Elapsed time	Fit-domain constr.	40	132	765
	3-index integrals	889	1575	3883
	Projection	44	64	227
	DF coefficients	18	66	331
	Assembly	202	393	1012
	Total DF	1181	2305	6427
	Total LMP2	2208	4935	11274
	E_2	−0.838846	−0.838878	−0.838866
	Local direct-space DF			
Elapsed time	Fit-domain constr.	39	309	844
	3-index integrals	958	1967	4656
	DF coefficients	221	319	517
	Assembly	64	77	190
	Total DF	1283	2674	6442
	Total LMP2	2553	5119	11500
	E_2	−0.838855	−0.838872	−0.838865

E_{HF} and E_2 are the HF and LMP2 energies per $(MgO)_3$ unit, respectively, in Hartree. The times are given in seconds. The VTZ fitting basis sets have been employed.

time is dominated by the 3-index integral evaluation, and particularly the two-electron integrals involving the GTO fitting functions, which constitute only a small part of the fitting basis. The major fraction of integrals involve the PTO fitting functions. These integrals are quite easy to compute because (see Section 2.2) they essentially reduce to one-electron integrals. Similar to the results of Table 2.4, the DRD and LD schemes demonstrate a comparable performance even for large unit cells. Those parts of the DRD calculation that cannot be reduced to linear scaling, e.g., the evaluation of the DF coefficients in the reciprocal space [41], are computationally insignificant even with large unit cells and thus do not spoil the general performance. However are still some problems with overhead, such as fit-domain generation and keeping track of the lists of integrals, coefficients, symmetry elements, etc., which in large unit cells can become quite cumbersome.

Another problem with large unit cells lies in memory management, especially when solving the LMP2 equations. In the present implementation, all the needed amplitudes are kept in memory, otherwise a permanent, excessive, and not sequential disk access would be needed [30]. Although the number of amplitudes also scales linearly by construction, in certain situations it still can exceed the available memory. A possible solution to this problem can be the application of the Laplace transformed MP2 formalism [1], which we have recently investigated in the context of the non-periodic LMP2 method [16].

Finally, as seen in Table 2.6, the stability of the energies obtained with different supercells, both in HF and LMP2 calculations, is very good, indicating that our implementation is indeed size extensive and can be safely applied to systems with relatively large unit cells.

2.4.6   Sodalite: A benchmark calculation

Silica polymorphs are a favorite test ground for the assessment of the quality of ab initio computational schemes. This is due both to their intrinsic importance and to the fact that reliable geometrical and relative stability data are experimentally available for this family of $(SiO_2)_n$ systems, which share the same basic unit, but exhibit an extraordinary variety of crystalline structures. Electron correlation appears to be an important factor affecting their stability order (see for instance [8]). Such systems are specially interesting due to the prospect of possible applications to zeolites, a class of microporous silica polymorphs that are important in many fields, such as catalysis, etc.

The following is a report on our preliminary LMP2 results for sodalite, the simplest of all zeolites, which demonstrates that the use of DF allows post-HF calculations of reasonable quality to be performed even for relatively complex crystalline systems.

The unit cell of sodalite is shown in Figure 2.2. Sodalite is a cubic crystal (space group $P\bar{4}3n$) with 36 atoms per cell and 48 symmetry operators. The geometry of our calculation is the experimental one [8].

The triple-zeta quality basis sets listed above are used for the Si and O atoms. There are 96 valence WFs in this structure, all localized on oxygen atoms.

The time needed for the HF calculation is roughly 2 hours on a single processor. This high performance is also due to the full exploitation of the high symmetry present in this structure. The time for localization is 20 hours, half of which is spent for the a posteriori symmetrization of the WFs [7]. This confirms that, at the present stage of implementation, the localization constitutes a bottleneck for large calculations.

Preliminary DF-LMP2 calculations have been performed with excitation domains, comprising the oxygen atom to which the WF belongs

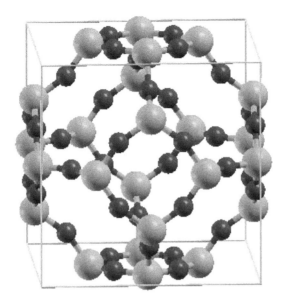

Figure 2.2 Sodalite structure.

as well as its two neighboring silicons. The LD-DF scheme with a VTZ fitting basis set, and 12-atom fit-domains has been employed. The over-all time of the LMP2 calculation was roughly 8 hours. About 2 hours are needed for the PAO generation, 3.5 hours for the evaluation of the three- and two-index DF integrals, approximately 2 hours for the other steps of the DF, and less than 1 hour is spent in the LMP2 equations. The correlation energy finally obtained amounts to -6.475136 Hartree per unit cell.

2.5   Conclusions

In this chapter we examined three different density fitting schemes for fac-torizing the electron repulsion integrals in periodic local MP2. They can be modified for the treatment of similar integrals that are needed in more advanced post-HF schemes. Test calculations were performed for a vari-ety of systems with small and medium-sized unit cells (up to 36 atoms), employing AO basis sets of triple-zeta quality. The related fitting basis sets primarily consist of (moment–less) Poisson basis sets, augmented by a few GTOs taking care of the multipole moments of the fitted orbital product densities.

The first DF scheme considered in this work is the local direct-space DF. It operates entirely in direct space and employs a priori restrictions in the fitting basis for individual orbital product densities (local fit-domains). In that sense it is similar to local fitting approaches proposed earlier for local correlation methods for molecular systems. Even though it does not exploit any reciprocal space techniques it is quite efficient even for rather small unit cells. The second DF scheme, multipole-corrected reciprocal-space DF, operates essentially in reciprocal space. The individual orbital product densities are partitioned into a partial density carrying the moments, and a momentless complement. The former is fitted in real space by GTOs, the latter in reciprocal space by Poisson-type orbitals. Unfortunately, this partitioning without further modification is not robust in Dunlap's sense [11], as pointed out in Section 2.3.2. Consequently, rather large fitting basis sets and fit-domains (in real space) are required in order to achieve sufficient accuracy, rendering this method in its present implementation as not competitive relative to the first. The last scheme, direct-reciprocal decoupled DF, can be considered as a further development of the second scheme. Again, one part of the density is fitted in reciprocal space by momentless Poisson-type orbitals, the complement in direct space by *projected GTOs* (the Poisson-type orbitals are projected out of the space spanned by the original GTOs). This method, in contrast to the second, is robust in Dunlap's sense. In contrast to the first it requires smaller fit-domains, but a k-mesh has to be specified. The first and the third schemes exhibit similar computational performance. The good performance of the first scheme for systems with small unit cells is indeed remarkable insofar as no reciprocal space techniques have been utilized. In any case, both schemes show close-to-linear scaling of the computational cost with unit cell size. This opens the path to an ab initio electron correlation treatment of systems with more complicated unit cells, comprising perhaps several dozen atoms. Several application studies utilizing the new code are presently underway.

Acknowledgment

The authors are grateful to the Deutsche Forschungsgemeinschaft for the financial support through an SPP 1145 grant.

References

[1] J. Almlöf. Elimination of energy denominators in Møller–Plesset perturbation theory by a Laplace transform approach. *Chem. Phys. Lett.*, 181:319, 1991.

[2] P. Y. Ayala, K. N. Kudin, and G. E. Scuseria. Atomic orbital Laplace transformed second-order Møller–Plesset theory for periodic systems. *J. Chem. Phys.*, 115:9698, 2001.

[3] E. J. Baerends, D. E. Ellis, and P. Ros. Self-consistent molecular Hartree–Fock-Slater calculations. I. The computational procedure. *Chem. Phys.*, 2:41, 1973.

[4] L. Belpassi, F. Tarantelli, A. Sgamellotti, and H. M. Quiney. Poisson-transformed density fitting in relativistic 4-component Dirac-Kohn-Sham theory. *J. Chem. Phys.*, 128:124108, 2008.

[5] J. W. Boughton and P. Pulay. Comparison of the Boys and Pipek-Mezey localizations and automatic virtual basis selection in the local correlation method. *J. Comput. Chem.*, 14:736, 1993.

[6] S. F. Boys and I Shavitt. A fundamental calculation of the energy surface for the system of three atoms. *University of Wisconsin, Report WIS-AF-13*, 1969.

[7] S. Casassa, C. M. Zicovich-Wilson, and C. Pisani. Symmetry-adapted Localized Wannier Functions suitable for periodic local correlation methods. *Theor. Chem. Acc.*, 116:726, 2006.

[8] B. Civalleri, C. M. Zicovich-Wilson, P. Ugliengo, V. R. Saunders, and R. Dovesi. A periodic ab initio study of the structure and relative stability of silica polymorphs. *Chem. Phys. Lett.*, 292:394, 1998.

[9] R. Dovesi, V. R. Saunders, C. Roetti, R. Orlando, C. Zicovich-Wilson, F. Pascale, B. Civalleri, K. Doll, N. M. Harrison, I. J. Bush, P. D'Arco, and M. Llunell. *CRYSTAL06 Users Manual*, 2006.

[10] B. I. Dunlap. Robust variational fitting: Gáspár's variational exchange can accurately be treated analytically. *J. Mol. Struct. (Theochem)*, 501–502:221, 2000.

[11] B. I. Dunlap, J. W. D. Connolly, and J. R. Sabin. On some approximations in applications of Xα theory. *J. Chem. Phys.*, 71:3396, 1979.

[12] C. Hättig and F. Weigend. CC2 excitation energy calculations on large molecules using the resolution of the identity approximation. *J. Chem. Phys.*, 113:5154, 2000.

[13] A. W. Hewat and C. Rieckel. The crystal structure of deuteroammonia between 2 and 180 K by neutron powder profile refinement. *Acta Cryst. A*, 35:569, 1979.

[14] A. F. Izmaylov and G. E. Scuseria. Resolution of the identity atomic orbital Laplace transformed second order Møller–Plesset theory for nonconducting periodic systems. *Phys. Chem. Chem. Phys.*, 10:3421, 2008.

[15] D. Kats, T. Korona, and M. Schütz. Local CC2 electronic excitation energies for large molecules with density-fitting. *J. Chem. Phys.*, 125:104106, 2006.

[16] D. Kats, D. Usvyat, and M. Schütz. On the use of the Laplace transform in local correlation methods. *Phys. Chem. Chem. Phys.*, 10:3430, 2008.

[17] W. Klopper and C. C. M. Samson. Explicitly correlated second-order Møller–Plesset methods with auxiliary basis sets. *J. Chem. Phys.*, 116:6397, 2002.

[18] R. Krishnan, J. S. Binkley, R. Seeger, and J. A. Pople. Self-Consistent Molecular Orbital Methods. XX. A Basis Set for Correlated Wavefunctions. *J. Chem. Phys.*, 72:650, 1980.

[19] M. Schütz, G. Hetzer, and H.-J. Werner. Low-order scaling local electron correlation methods. I. linear scaling local MP2. *J. Chem. Phys.*, 111:5691, 1999.

[20] F. R. Manby, D. Alfe, and M. J. Gillan. Extension of molecular electronic structure methods to the solid state: computation of the cohesive energy of lithium hydride. *Phys. Chem. Chem. Phys.*, 8:5178, 2006.

[21] F. R. Manby, P. J. Knowles, and A. W. Lloyd. The Poisson equation in density fitting for the Kohn-Sham Coulomb problem. *J. Chem. Phys.*, 115:9144, 2001.

[22] F. R. Manby and P. J. Knowles. The Poisson equation in the Kohn-Sham Coulomb problem. *Phys. Rev. Lett.*, 87:163001, 2001.

[23] M. Marsman, A. Grüneis, J. Paier, and G. Kresse. Second-order Møller–Plesset perturbation theory applied to extended systems. I. Within the projector-augmented-wave formalism using a plane wave basis set. *J. Chem. Phys.*, 130:184103, 2009.

[24] L. Maschio and D. Usvyat. Fitting of local densities in periodic systems. *Phys. Rev. B*, 78:073102, 2008.

[25] L. Maschio, D. Usvyat, F. Manby, S. Casassa, C. Pisani, and M. Schütz. Fast local-MP2 method with Density-Fitting for crystals. A. Theory. *Phys. Rev. B*, 76:075101, 2007.

[26] J. W. Mintmire and B. I. Dunlap. Fitting the Coulomb potential variationally in linear-combination-of-atomic-orbitals density-functional calculations. *Phys. Rev. A*, 25:88, 1982.

[27] F. Pascale, C. M. Zicovich-Wilson, R. Orlando, C. Roetti, P. Ugliengo, and R. Dovesi. Vibration Frequencies of $Mg_3Al_2Si_3O_12$ Pyrope. An ab Initio Study with the CRYSTAL Code. *J. Phys. Chem.*, 109:6146, 2005.

[28] B. Paulus. The method of increments — a wavefunction-based ab initio correlation method for solid. *Phys. Rep.*, 428:1, 2006.

[29] C. Pisani, M. Busso, G. Capecchi, S. Casassa, R. Dovesi, L. Maschio, C. Zicovich-Wilson, and M. Schütz. Local-MP2 electron correlation method for non conducting crystals. *J. Chem. Phys.*, 122:094113, 2005.

[30] C. Pisani, L. Maschio, S. Casassa, M. Halo, M. Schütz, and D. Usvyat. Periodic Local MP2 Method for the Study of Electronic Correlation in Crystals: Theory and Preliminary Applications. *J. Comp. Chem.*, 29:2113, 2008.

[31] R. Polly, H.-J. Werner, F. R. Manby, and P. J. Knowles. Fast Hartree–Fock theory using local density fitting approximations. *Mol. Phys.*, 102:2311, 2004.

[32] S. Saebø and P. Pulay. Local treatment of electron correlation. *Annu. Rev. Phys. Chem.*, 44:213, 1993.

[33] D. M. Schrader and S. Prager. Use of electrostatic variation principles in molecular energy calculations. *J. Chem. Phys.*, 37:1456, 1962.

[34] M. Schütz. Low-order scaling local electron correlation methods. III. linear scaling local perturbative triples correction (T). *J. Chem. Phys.*, 113:9986, 2000.

[35] M. Schütz and F.R. Manby. Linear scaling local coupled cluster theory with density fitting I: 4-external integrals. *Phys. Chem. Chem. Phys.*, 5:3349, 2003.

[36] M. Schütz, H.-J. Werner, R. Lindh, and F.R. Manby. Analytical energy gradients for local second-order møller-plesset perturbation theory using density fitting approximations. *J. Chem. Phys.*, 121:737, 2004.

[37] H. Stoll. Correlation energy of diamond. *Phys. Rev. B*, 46:6700, 1992.

[38] G. Stollhoff and P. Fulde. A local approach to the computation of correlation energies of molecules. *Z. Phys. B*, 26:251, 1977.

[39] G. te Velde and E. J. Baerends. Precise density-functional method for periodic structures. *Phys. Rev. B*, 44:7888, 1991.

[40] S. B. Trickey, J. A. Alford, and J. C. Boettger. Methods and Implementation of Robust, High-precision Gaussian Basis DFT Calculations for Periodic Systems: the GTOFF Code. *Computational Materials Science, Theoretical and Computational Chemistry series, Vol. 15, edited by J. Leszczynski,*, page 171, Elsevier, 2004.

[41] D. Usvyat, L. Maschio, F. Manby, S. Casassa, M. Schütz, and C. Pisani. Fast local-MP2 method with Density-Fitting for crystals. B. Test calculations and application to the carbon dioxide crystal. *Phys. Rev. B*, 76:075102, 2007.

[42] D. Usvyat, L. Maschio, and M. Schütz. On the combined use of Gaussian- and Poisson-type orbitals as auxiliary functions for periodic density fitting. *to be submitted*, 2010.

[43] O. Vahtras, J. Almlöf, and M. W. Feyereisen. Integral approximations for LCAO-SCF calcualtions. *Chem. Phys. Lett.*, 213:514, 1993.

[44] L. Valenzano, Y. Noel, R. Orlando, C. M. Zicovich-Wilson, M. Ferrero, and R. Dovesi. Ab Initio vibrational spectra and dielectric properties of carbonates: Magnesite, Calcite and Dolomite. *Theor. Chem. Acc.*, 117:991, 2007.

[45] S. Varga. Long-range analysis of density fitting in extended systems. *Int. J. Quantum Chem.*, 108:1518, 2008.

[46] S. Varga, M. Milko, and J. Noga. Density fitting of two-electron integrals in extended systems with translational periodicity: the Coulomb problem. *J. Chem. Phys.*, 124:034106, 2006.

[47] F. Weigend, M. Häser, H. Patzelt, and R. Ahlrichs. RI-MP2: optimized auxiliary basis sets and demonstration of efficiency. *Chem. Phys. Lett.*, 294:143, 1998.

[48] H.-J. Werner, P.J. Knowles, and F.R. Manby. Fast linear scaling second-order Møller–Plesset perturbation theory (MP2) using local and density fitting approximations. *J. Chem. Phys.*, 118:8149, 2003.

[49] C. M. Zicovich-Wilson. Two points of view to look at symmetry. *J. Phys: Conf. Series*, 117:012030, 2008.

chapter three

The method of increments— a wavefunction-based correlation method for extended systems

Beate Paulus and Hermann Stoll

Contents

This chapter focuses on the method of increments, which is based on a many-body expansion of the correlation energy of the solid in terms of contributions from finite numbers of localized orbitals, and which allows for accurate wavefunction-based methods (at the coupled-cluster level, e.g.) to be used for evaluating individual contributions. The method of increments has been applied to a great variety of materials, from covalent semiconductors to ionic insulators, from large band-gap materials like diamond to the half-metal α-tin, from large molecules like fullerenes over polymers and graphite to three-dimensional solids. Rare-gas crystals, where the bonding is van der Waals–like, are treated as well as group 2 and 12 metals where the metallic bonding is mostly due to correlation. A recent development, which is presented in this chapter, is the application of the method of increments to the adsorption energies of molecules on surfaces.

3.1   Introduction

All ab initio electronic structure methods serve the purpose of deriving properties of a physical system from the corresponding electronic Hamiltonian. This task is made difficult by the fact that the interelectronic interaction in the Hamiltonian involves two-body operators, which can be treated only approximately within a mean-field approach.

One very successful approximation is density-functional theory (DFT) [1–4], which relies on the ground-state density of the system and avoids calculation of its many-body wavefunction. The density is generated within a (formally) mean-field scheme, and all the difficulties are hidden in a (universal) exchange-correlation functional of the electron density. Successful local and semilocal approximations to this functional are known, but it is difficult to improve on them in a systematic way.

An alternative approach with the advantage of the possibility of systematical improvement is constituted by wavefunction-based methods. Here, the starting point is the mean-field approach introduced by Hartree [5], Fock [6], and Slater [7]. The Hartree–Fock wavefunction, which has the form of a Slater determinant built from one-particle orbitals, can then be used for subsequent correlation calculations. Most of the quantum-chemical correlation methods rely on an Ansatz for the many-body wavefunctions in terms of configurations. In the configuration interaction (CI) Ansatz [8, 9], e.g., determinants are considered where one, two, or more orbitals occupied in the ground-state Slater determinant are replaced by unoccupied ones. The weights of these determinants in the Ansatz for the many-body wavefunction are optimized by minimizing the total energy of the system.

In order to apply these types of quantum-chemical correlation methods to an extended system, it is essential to use the property that the correlation hole (involving interelectronic distances, for which the electron interactions must be treated beyond the mean-field level) is fairly local. It is useful for a solid, therefore, to switch from the description with delocalized Bloch orbitals to localized Wannier orbitals [10]. The reformulation of the many-body wavefunction in terms of localized orbitals triggers the group of the so-called *local correlation methods*, e.g., the local Ansatz developed by Stollhoff and Fulde [11–13], or the local correlation method by Pulay [14] with its modification to periodic systems (CRYSCOR project [15–24]; see also Chapter 2). In these methods, excitations from pairs of localized orbitals with distances beyond a certain threshold are neglected. But even if the correlation hole extends only to second-nearest neighbors around a given localized orbital, that corresponds to 4 times 24 orbital pairs per unit cell in the case of diamond, for example. It is hard to correlate so many orbital pairs simultaneously with highly accurate wavefunction-based correlation methods.

Therefore, further approximations may be required. To this purpose the method of increments was developed [25–27]. In this approach the correlation energy is calculated, in the first step, as a sum of independent correlation contributions of localized orbitals (or groups of localized orbitals). As corrections, the nonadditive parts coming from two, three, or more (groups of) localized orbitals are added. Thus, the number of orbitals to be correlated simultaneously can be kept small. If the group of localized orbitals is chosen reasonably (so as to minimize intergroup correlation) and the correlation method applied is size-extensive, this partitioning of the correlation energy of the solid yields a rapidly converging series.

In the local Ansatz and in the local correlation method by Pulay, the virtual space is truncated for each pair of occupied orbitals, to keep the computational effort manageable. In the method of increments, an a priori truncation of the virtual space is usually not necessary, for the following reason. The number of basis functions needed to evaluate a given increment is small, since the number of orbitals to be correlated simultaneously is small; additional basis functions are only needed insofar as they carry the Hartree–Fock field of the surroundings, and if the Hartree–Fock field is suitably approximated, the number of additional basis functions can be kept small. But of course a combination of the method of increments with the local correlation method of Pulay would be possible to reduce the computational effort even further [28, 29].

In the last decades the method of increments was applied to ground-state properties of various material classes: from insulators [30–34] over semiconductors [25, 26, 35–39] to metals [27, 40–50], from strongly bound ionic or covalent systems to weakly bound van der Waals solids [51–53], from large molecules [28, 54] over polymers [55–62] to three-dimensional solids, from weakly correlated systems to strongly correlated ones such as transition-metal oxides [63, 64] and rare-earth nitrides and oxides [65–68]. In all cases the obtained ground-state properties agree well with the experimental values. A review on the performance of the method of increments for systems with a band gap can be found in [69].

This chapter discusses two recent extensions of the method of increments: to metals, which need a special treatment due to the vanishing energy gap at the Fermi level and the difficulties in the localization of the orbitals (see also [70]), and to the calculation of adsorption energies of molecules on surfaces. The physisorbed molecules are of interest especially in the second case, because their binding is dominated by dispersion forces. It is known that standard density functionals are not well suited for describing this kind of interaction.

3.2   The method of increments

3.2.1   General ideas

For applying the method of increments, one divides the total energy of the system into a Hartree–Fock part, which is always directly calculated for the extended system (e.g., with the program code CRYSTAL [71]), and a correlation part. The correlation energy is determined by means of the incremental scheme. To exploit the fact that the correlation effects are short-range, it is reasonable to transform the extended Bloch orbitals of the periodic system (or the delocalized canonical orbitals in a finite but extended system) to localized orbitals. In the case of a periodic system, these orbitals are called *Wannier orbitals*; in the case of finite systems, various names are used according to the different localization procedures in use (e.g., Foster–Boys [72] or Pipek–Mezey [73]). With these orbitals it is possible to divide the occupied space into disjunct localized orbital groups whose periodic repetition will generate the whole occupied orbital space of the crystal.

The method of increments uses these orbital groups to expand the correlation energy of the system in terms of local contributions (for a sketch see Figure 3.1). It is formally similar to treating the hierarchy of nth-order atomic Bethe–Goldstone equations [74]. Here we want to sketch the basic ideas and some important formulae (for more details see [69] and [75]).

In a first step, the one-body correlation-energy increments ϵ_i are obtained by correlating each of the localized orbital groups separately while keeping the other ones frozen at the Hartree–Fock level. Any size-extensive correlation method can be used for this calculation. (We usually use the coupled-cluster approach with single and double excitations and perturbative triples (CCSD(T), see, e.g., [77] and [78]).) This yields a first

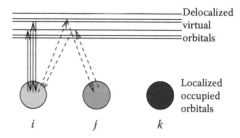

Figure 3.1 Schematic overview of the method of increments using localized orbital groups i, j, k in the occupied Hartree–Fock space and excitations into the whole virtual space.

approximation of the correlation energy per unit cell

$$E_{\text{corr}}^{(1)} = \sum_{i \in \text{u.c.}} \epsilon_i, \tag{3.1}$$

which corresponds to the correlation energy of independent orbital groups (independent pair approximation).

In the next step we include the correlations of pairs of orbital groups i, j, where i labels the orbital groups in the unit cell and j all orbital groups of the system. As a correction to our first approximation of the correlation energy of the whole system, $E_{\text{corr}}^{(1)}$, only the nonadditive part $\Delta\epsilon_{ij}$ of the two-body correlation energy ϵ_{ij} is needed.

$$\Delta\epsilon_{ij} = \epsilon_{ij} - (\epsilon_i + \epsilon_j). \tag{3.2}$$

Higher-order increments are defined analogously, e.g., the three-body increment

$$\Delta\epsilon_{ijk} = \epsilon_{ijk} - (\epsilon_i + \epsilon_j + \epsilon_k) - (\Delta\epsilon_{ij} + \Delta\epsilon_{jk} + \Delta\epsilon_{ik}). \tag{3.3}$$

The correlation energy of the system is finally obtained by adding up all the increments with appropriate weight factors:

$$E_{\text{corr}} = \sum_{i \in \text{u.c.}} \epsilon_i + \frac{1}{2} \sum_{\substack{i \in \text{u.c.} \\ j \in \text{crystal} \\ i \neq j}} \Delta\epsilon_{ij} + \frac{1}{6} \sum_{\substack{i \in \text{u.c.} \\ jk \in \text{crystal} \\ i \neq j \neq k}} \Delta\epsilon_{ijk} + \cdots . \tag{3.4}$$

It is obvious that by calculating higher and higher increments the exact correlation energy should be reached, within the correlation method applied, provided the expansion converges at all.

However, more than mere convergence is required. The partitioning of the correlation energy only makes sense if the series converges quickly enough, both with the order of increments (i.e., three-body increments should be significantly smaller than two-body increments, etc.) and with the distance of the orbital groups involved in the increments. Increments involving distant orbital groups must decay faster than the number of increments increases in a 3-dimensional system. For a semiconductor or an insulator, increments with distant orbital groups show a van der Waals-like decay $\sim r^{-6}$, whereas the number of pairs grows with r^2, so an overall decay of r^{-4} guarantees convergence with respect to the distance of orbital groups.

The actual calculation of the increments is performed in finite embedded fragments of the crystal under consideration. This means that the orbitals to be correlated are surrounded by frozen Hartree–Fock orbitals (possibly described with a lower-level basis set), which in turn are embedded in a field of point charges or saturated by terminating hydrogens.

This allows us to use standard quantum chemical programs with their standard localization procedures and various options for correlation treatment. (For all our calculations so far, we used the program package MOLPRO [76].) We test the quality of our embedding and the size of the fragments chosen as follows. We perform a series of calculations for the same increment (mostly the one-body increments and the largest two-body increments in the system) in fragments of increasing size and improved embedding. We select fragments where the differences between corresponding increments are converged below a chosen threshold. In systems with a band gap we can reduce the error due to the finite fragments below 3% of the correlation contribution to the cohesive energy.

A fully automated implementation of the incremental scheme has been developed for finite extended systems [79], including an error analysis [80], energy screening [81], the use of domain-specific basis sets [82], an automatized procedure for using molecular symmetry [83], and including corrections for core–core and core–valence correlations [84]. This automated procedure was applied, for example, to hydrocarbons, glycine oligomers [82], and water clusters [85, 86].

3.2.2   Extension to metals

A direct transfer of the incremental approach to metallic systems is not possible since localized orbitals become very long-range entities, with an algebraic rather than exponential decay. This will slow down the convergence of the incremental expansion. But even if we assume that convergence will be reached eventually, the fragments of the crystal needed for correlating increments will be extremely large. In addition, the form of the localized orbitals in metals is sensitive to the localization criterion and the fragment in which they are generated, i.e., it is extremely difficult to generate fragments of the crystal whose localized orbitals correspond to those of the full system.

In order to make the expansion still computationally feasible, we have suggested [41, 54] starting from suitable nonmetallic model systems where long-range orbital tails are absent. For group 2 or 12 metals, e.g., we leave out p functions in the first stage, which leads to a system of nonbonded s-type atoms. The transition from this model system to the physical one is described step by step within the incremental expansion. Thus, the basis set for the group 2 or 12 atoms is extended for one, two, three, or more atoms at a time. This way, delocalization effects are successively accounted for in the course of the expansion. More specifically, when calculating pair contributions for a given orbital group combination (i, j), we allow for delocalization $i \rightarrow j$ and $j \rightarrow i$, and similarly with the 3-body terms we allow for delocalization over the triples of atoms, etc. It is clear that the final result is not affected; only the convergence properties of the many-body expansion are changed. As an additional advantage, we can calculate individual terms of the expansion from finite fragments of the crystal of reasonable size (comprising the atoms to be correlated embedded in a surrounding of frozen "model atoms"). A schematic overview of the whole procedure is given in Figures 3.2 and 3.3.

3.2.3   Extension to surface adsorption

In the description of the method of increments so far, the many-body expansion was applied to the total correlation energy. In the case of adsorption of a molecule on a surface, we want to determine the adsorption energy, or in other words, the interaction energy of the molecule with the substrate. To quantify the molecule–surface interaction we define the *adsorption energy* as

$$E_{ads} = E_{mol/surf} - E_{surf+ghosts} - E_{mol+ghosts} + \Delta E_{mol} + \Delta E_{surf} \quad (3.5)$$
$$= E_{ads}^{HF} + E_{ads}^{corr} . \quad (3.6)$$

$E_{mol/surf}$ is the total energy of the joint system, and $E_{surf+ghosts}$ and $E_{mol+ghosts}$ are the energies of the fragments at the same geometry as in the joint system, using the same basis sets (that of the joint system) in all cases. By this means, we correct for the basis set superposition error (BSSE) according the counterpoise scheme of Boys and Bernardi [87]. ΔE_{mol} and ΔE_{surf} are the energy changes induced by the relaxation of the adsorbed molecule in the gas phase, and that of the free surface. Neglecting the latter two terms leads to significant simplifications (cancellations of terms) in the many-body incremental expansion. However, the adsorption energy is then, in principle, not totally adequate for comparison with the experiment. In any event, for physisorption the structure relaxation is usually small and the adsorption energy should not be much affected.

If the method of increments is applied to the calculation of the correlation contribution to the adsorption energies [88–92] E_{ads}^{corr}, the natural choice of the localized orbital groups are the molecule (labeled M) and localized orbital groups within the surface (labeled $i, j, k, \ldots$); see Figure 3.4. It is only necessary to evaluate all increments that occur or change due to the adsorption of the molecule on the surface. For example, the one-center contributions from the adsorbed molecule and the substrate, respectively, are

$$\eta_M = \epsilon_M^{mol/surf} - \epsilon_M^{free} \quad (3.7)$$

and

$$\eta_i = \epsilon_i^{mol/surf} - \epsilon_i^{surf}. \quad (3.8)$$

Other contributions occur in the adsorption system due to the simultaneous correlation of orbitals involving groups from the molecule and the surface,

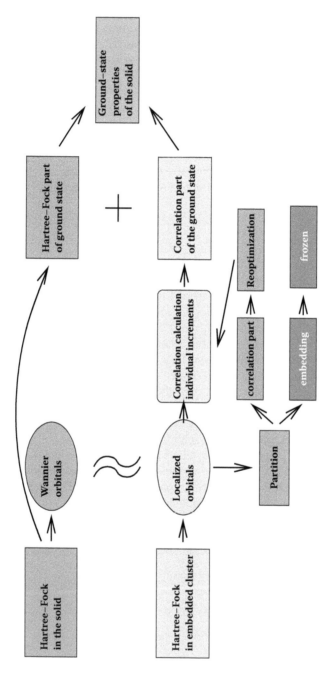

Figure 3.2 Schematic overview of the procedure of the ground-state calculations with the method of increments.

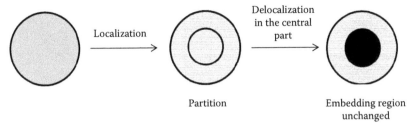

Figure 3.3 Schematic overview of the localization procedure in metals.

e.g., η_{Mi} and η_{Mij}. It is clear that increments for localized orbital groups in the surface, like ϵ_i or ϵ_{ij}, which are sufficiently far away from the adsorbed molecule, cancel out, i.e., $\eta_i = 0$ or $\eta_{ij} = 0$. E_{ads}^{corr} can now be calculated as the sum of all η terms,

$$E_{ads}^{corr} = \eta_M + \sum_i \eta_i + \sum_{\substack{ij \\ i \neq j}} \eta_{ij} + \sum_i \eta_{Mi} + \cdots . \tag{3.9}$$

An advantage of the method of increments (in addition to the increased computational speed) is that we have access to the individual contributions from the different groups/atoms and can discuss their impact on binding.

3.3 Applications

3.3.1 Application to systems with a band gap

Since the first formulation of the method of increments as a many-body expansion of the correlation energy of periodic systems in terms of local correlation energy increments [25–27], the method has been applied to a great variety of periodic systems with a band gap. These applications have been reviewed in [69]. Here we want to briefly sketch results for various ground-state properties: In Figure 3.5 we summarize results for cohesive energies of ionic crystals (left panel) and semiconductors (right panel). In both cases

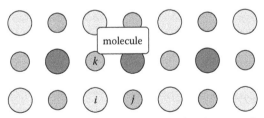

Figure 3.4 Schematic overview of the localized orbital groups for the adsorption on an ionic surface.

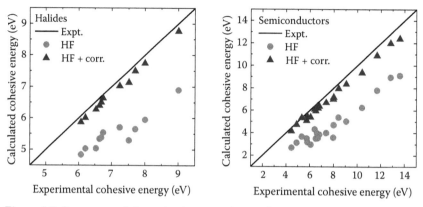

Figure 3.5 Summary of the cohesive energies at the Hartree–Fock level and at the correlated level using the method of increments for various ionic materials (left panel) and semiconductors (right panel). (From Voloshina, E., and B. Paulus. *Chemical Modelling: Applications and Theory*, Vol. 6, M. Springborg (ed.). RSC, Cambridge, 2009.)

we compare the Hartree–Fock cohesive energy and the correlated cohesive energy determined at the coupled cluster level with the experimental one. At the Hartree–Fock level of theory we obtain between 50 and 80% of the experimental value, and including electron correlation, we recover more than 95% for all substances (13 alkali-halides and alkaline-earth oxides, 20 semiconductors). The missing part is mainly due to the limited atomic basis set, which was of valence-triple-zeta quality in most cases. Consider as a special example germanium, a typical semiconductor. Whereas at the Hartree–Fock level the cohesive energy is only 53% of the experimental value, including correlations we achieve 95%. Therefore, nearly half of the cohesive energy is due to electron correlation, even though the material would not be regarded as strongly correlated.

Even more pronounced is the role of correlation in the case of the rare-gas crystals. At very low temperatures the rare gas crystals crystallize in a dense-packed face-centered cubic (fcc) structure, and the attraction between the closed-shell atoms is entirely due to van der Waals interactions. The ab initio description of rare gas crystals is a challenging problem of computational physics. The weak van der Waals interaction can only be described by very accurate methods—highly correlated methods like coupled cluster theory and very good basis sets are necessary for calculating the ground-state properties of such crystals. A Hartree–Fock description does not yield binding at all. In the original formulation of the method of increments, the Hartree–Fock contribution to the ground-state energy of the crystal was taken from fully periodic calculations, and a many-body expansion in terms of local increments was performed for electron-correlation contributions only. In the case of rare gas crystals, where binding is due

Table 3.1 Cohesive Energies (in μH) for the fcc
Structure of Rare Gas Crystals (Calculated at the
Experimental Lattice Constants), Optimized Lattice
Constants (in Å), and Bulk Moduli (in kBar), in
Comparison to Experimental Values

	Ne	Ar	Kr	Xe
E_{cal}	−750.9	−2943.0	−4263.9	−6052.0
E_{exp}	−752.2	−2943.9	−4264.3	−6051.2
a_{cal}	4.468	5.311	5.633	6.111
a_{exp}	4.464	5.311	5.670	6.132
B_{cal}	10.4	27.9	32.9	37.2
B_{exp}	10.9	26.7	36.1	36.4

Calculated values include electronic contributions up to 4-body
terms at the CCSD(T) level and the zero-point-energy determined from the phonon dispersion relation. For details and references, see K. Rościszewski, B. Paulus, P. Fulde, and H. Stoll,
Phys. Rev. B 60, 7905 (1999). [51]

to correlation effects and no long-range HF contributions are present, we
expand the total ground-state energy as a whole. Therefore, the one-body
contribution cancels out for the cohesive energy, and the incremental series
starts with the two-body contributions. It is crucial for accurate results to
correct each increment for the basis set superposition error using the counterpoise correction. We summarize the calculated ground-state properties,
i.e., cohesive energy, lattice constant, and bulk modulus of the rare-gas
crystals neon to xenon in Table 3.1.

For all three ground-state properties and for all four rare gas crystals
we obtain excellent agreement with the experimental values. In the case
of neon it is crucial to include the zero-point energy—it contributes about
30% to the cohesive energy. In the case of the heavier rare gas crystals, the
relativistic effects are important. We include them via a scalar-relativistic
pseudopotential. Additionally, the three-body contributions become more
important for the heavier rare gas crystals: for example, for xenon they
reduce the cohesive energy by about 8%. Only a detailed evaluation of all
the individual contributions can yield such an excellent agreement with the
experiment. Only the partitioning of the correlation energy provided by the
method of increments makes it feasible to calculate all these contributions
at a very high level of accuracy with reasonable computational effort.

As a third class of materials, we discuss results with the method of
increments for carbon-based systems with delocalized π electrons. Our applications include graphene [27], fullerene [54], polyacetylene [55,56], and
poly-paraphenylene [60]. In this class of materials the problem lies in the localization of the orbitals. In all cases we made a separation into the σ-bond
system, which is well localized, and into the π-electrons, which form only

Table 3.2 The Cohesive Energy per Carbon Atom (in Hartree), Evaluated with Different Basis Sets (Valence Double-Zeta and Triple-Zeta) at the Hartree–Fock and Correlated Levels, is Listed for Different Carbon Materials

energy/atom	HF	HF + corr vdz	HF + corr vtz	exp
C_{60}	−0.189	−0.242	−0.253	—
Diamond	−0.197	−0.254	−0.263	−0.277
Graphene	−0.203	−0.253	—	−0.277

For details and references, see H. Stoll, *J. Chem. Phys.* 97, 8449 (1992); B. Paulus, P. Fulde, and H. Stoll, *Phys. Rev. B* 51, 10572 (1995); and B. Paulus, *Int. J. of Quant. Chem.* 100, 1026 (2004).

modestly well-localized orbitals centered around every second σ-bond. Here, we present calculations for the pure carbon-based materials graphene (infinite two-dimensional system) and the fullerene molecule, C_{60}, in comparison to diamond, see Table 3.2. The Hartree–Fock contribution is about 70% of the experimental value for diamond or graphite. There exists no experimental value for C_{60}, but it is not expected to be much different from diamond or graphite. The correlation contributions to cohesion (with a double-zeta basis set) are largest for diamond and smallest for graphite. The result for C_{60} is placed nicely between these two values, showing the duality of the bonding structure of the fullerene molecule between those of diamond and graphite. It is expected from the concept of electron correlation, that more strongly localized electrons like those in the σ-bonds of diamond are subjected to larger correlation effects than the delocalized π-electrons of graphene. Overall we reach about 92% of the cohesive energy for diamond or graphite with the double-zeta basis set and about 96% with the triple-zeta basis. The main part of the missing contributions is very probably due to shortcomings of the finite basis set rather than to the truncation of the incremental expansion. This application shows that it is also possible to apply a local correlation scheme like the method of increments to delocalized π electron systems without any significant loss of accuracy. Of course the decay of the two-body increments with distance within the π orbitals is slower than for the σ orbitals, but still manageable with reasonable computational effort.

3.3.2 Application to group 2 and 12 metals

Density-functional approaches are expected to work well for metals because the functionals used are mainly based on the local-density approximation of the homogeneous electron gas or gradient-corrected versions of it. This is especially true for metals with nearly homogeneous distribution of the electron density like sodium, aluminium, and magnesium. However, a strong dependence on the functional has been found for ground-state properties of zinc, cadmium [93], and mercury [43]. Zinc and cadmium crystallize in a hexagonal close-packed (hcp) structure with an anomalously large c/a ratio (the bond length in c direction is about 10% larger than within the hexagonal plane). Whereas all functionals yield reasonable results for the in-plane bond distance, results for the lattice parameter c vary within 1 Å depending on the functional. For mercury the situation is even worse. Mercury crystallizes below $-40°C$ into a rhombohedral structure with an angle of 70°. Different functionals yield angles of between 90°, which corresponds to a simple cubic lattice, and 60°, which corresponds to a face-centered cubic lattice. These results are not at all satisfactory. One would like to have a systematically improvable ab initio method for such materials. As discussed in Section 3.2.2, wavefunction-based ab initio methods can be applied here when using an incremental scheme starting from a localizable model system. The starting-point is essentially a superposition of closed-shell atoms, for which it is easy to generate localized orbitals. In the course of the incremental expansion, the orbital spaces, for which delocalization and correlation are taken into account, are embedded into surroundings of frozen localized orbitals of the model system [46]. So far we have applied this embedding scheme to magnesium [45], zinc [47,49], cadmium [47,50], and mercury [42,43], and tested the convergence carefully with respect to the embedding properties, the distance of the localized orbital groups, and the order of increments. Details can be found in the publications cited previously. Here, we only summarize the results.

In Table 3.3 the individual contributions to the cohesive energies are listed. The Hartree–Fock treatment yields no binding for the group 12 metals, and only 20% of the experimental cohesive energy for magnesium.

Table 3.3 The Cohesive Energies per Atom (eV) of Different Group 2/12 Metals

E_{coh}	Mg	Zn	Cd	Hg
HF	−0.27	+0.09	+0.25	+0.98
One-body	+0.1019	−0.0302	−0.0026	+0.1170
Two-body	−1.3646	−1.4517	−1.5248	−1.4760
Three-body	+0.0580	+0.0344	+0.0974	−0.2740
Four-body	+0.0153	+0.0002	+0.0002	+0.0820
Total	−1.46	−1.36	−1.18	−0.57
Exp	−1.50	−1.35	−1.16	−0.67

Energies were evaluated at the various stages of the incremental expansion, using CCSD(T) calculations with triple-zeta basis sets. Experimental values have been corrected by zero-point energies. For details and references, see B. Paulus, K. Rościscewski, N. Gaston, P. Schwerdtfeger, and H. Stoll, *Phys. Rev. B* 70, 165106 (2004); E. Voloshina and B. Paulus, *Phys. Rev. B* 75, 245117 (2007); and N. Gaston and B. Paulus, *Phys. Rev. B* 76, 214116 (2007).

Including electron correlation at the CCSD(T) level with valence triple-zeta basis sets, we achieve very good agreement with the experiment in all cases. The main contribution to binding is due to the two-body contributions, whereas the one-body part and higher-order increments are repulsive. An exception is mercury, where the sum over the three-body increments is also attractive (as for the two-body terms), possibly due to the different lattice structure. It is worth mentioning that for mercury about half of the correlation contribution to binding is due to the correlation of the closed 5*d* shell. This means that neglecting outer-core contributions would yield mercury still unbound. In addition to the calculations discussed so far, for which we applied small-core relativistic pseudopotentials, we also performed calculations, where we used a nonrelativistic pseudopotential for mercury. In that case, mercury turns out to be overbound by a factor of two compared to the experiment, the cohesive energy becomes comparable to Cd, and one would expect a much higher melting point. This gives a hint that relativistic effects are probably responsible for the low melting temperature of mercury.

We not only determined the cohesive energy, we also used the incremental scheme to calculate the lattice parameters of magnesium, zinc, cadmium, and mercury, and the corresponding bulk moduli. In the case of magnesium, we could significantly improve on the Hartree–Fock results and achieve good agreement with the experiment, which proves that the method of increments will also work for quasi-free electron metals. Even more interesting results were determined for mercury, zinc, and cadmium. In the case of mercury (see Table 3.4), the local density approximation (LDA) functional more or less fortuitously yields the correct rhombohedral structure but a bulk modulus with an error of 50%. The method of increments already yields the rhombohedral structure at the two-body level, but in order to achieve a quantitatively satisfactory result for the bulk modulus, the inclusion of three-body increments is crucial.

Table 3.4 The Lattice Parameters of the Rhombohedral Ground-State Structure of Solid Mercury at Different Levels of Theory Compared to the Experimental Values

Method	$a(\text{Å})$	$\alpha(°)$	B(Mbar)
LDA	2.97	72.6	0.176
HF + 1-body + 2-body-corr.	2.97	70.0	0.132
+3-body: s-only	2.97	69.7	0.383
+3-body: sd-corr	2.96	69.5	0.360
Expt.	3.005	70.53	0.382/0.322

For details and references, see N. Gaston, B. Paulus, K. Rościscewski, P. Schwerdtfeger, and H. Stoll, *Phys. Rev. B* 74, 094102 (2006).

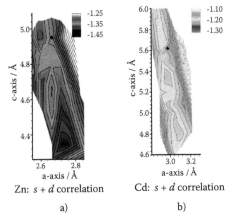

Figure 3.6 The potential energy surfaces for zinc (a) and cadmium (b), in dependence on the lattice parameters a and c of the hcp structure. The energy was calculated at the CCSD(T) level applying the method of increments. The diamond indicates the experimental structure. Data from Gaston, N., et al. *Phys. Chem. Chem. Phys.*, 12, 681 (2010), with permission of the Royal Society of Chemistry.

We also determined the lattice structure for zinc and cadmium, in order to get an insight into the bonding of these materials and to detect the reason for the anomalously large c/a ratio. The potential energy surfaces depending on the two lattice parameters a and c are plotted in Figure 3.6. In both cases, we could reproduce not only the experiment-like minimum with the large c/a ratio, but also found a minimum at larger a values and drastically smaller c values, which has a nearly ideal hcp structure (see Table 3.5). This ideal-like minimum is much more pronounced in zinc, where a ridge shows up in the potential energy surface separating the experimental and the ideal-like minimum. High-pressure experiments cannot enlarge the a values and therefore stay on one side of the ridge, finding a third minimum (min-3) with a similar small a value as the first one, but also a smaller c value. This would correspond to a special symmetric point in the hcp lattice structure, both in real and reciprocal space. The situation is somewhat different in cadmium. Again we have the two minima, but in cadmium the ridge is lower in the c direction, separating the experiment-like minimum from a flat region extending from approx. $a = 3.0$, $c = 5.4$ Å, to $a = 3.1$, $c = 4.8$ Å. This region can also be found with DFT, or with an incremental CCSD(T) treatment neglecting d correlations and three-body terms. However, we find that both in zinc and cadmium we can fully resolve the complicated structure of the potential energy surface of the group 12 hcp metals only when properly treating the outer-core d correlation.

Table 3.5 Lattice Constants (Å), Cohesive Energies (eV), and Cell Volumes (Å^3) for Minima on the Potential Energy Surface PES(a,c) of Zinc and Cadmium

		a	c	c/a	E_{coh}	$V_{u.c.}$
Zinc	Expt-like	2.61	4.98	1.91	−1.35	29.3
	Ideal-like	2.72	4.34	1.60	−1.39	27.8
	Min-3	2.61	4.70	1.80	−1.35	27.7
	Experiment	2.67	4.95	1.86	−1.35	30.4
Cadmium	Expt-like	2.92	5.61	1.92	−1.19	41.4
	Ideal-like	3.1	4.8	1.55	−1.20	
	Experiment	2.98	5.62	1.89	−1.16	43.2

For details and references, see N. Gaston, B. Paulus, U. Wedig, and M. Jansen, *Phys. Rev. Lett.* 100, 226404 (2008); N. Gaston, D. Andrae, B. Paulus, U.Wedig, and M. Jansen, *Phys. Chem. Chem. Phys.*, submitted (2009).

3.3.3 *Application to adsorption on CeO$_2$ and graphene*

As described in the methodological section, it is possible to extend the method of increments to calculating adsorption energies of molecules on surfaces. We tested the incremental expansion for two different molecules (CO and N_2O) on two different surfaces of CeO_2 [88–90]. There is significant interest in accurate values of adsorption energies and in understanding the character of the bonding of these small molecules physisorbed on ceria. This is so because it is an initial process in today's car catalysts, where it influences the catalytic oxidation reaction of CO to form CO_2. Due to the quite weak interaction between molecule and surface, which is dominated by two types of interaction, namely dispersion and/or classical electrostatics, highly accurate wavefunction-based correlation methods like MP2 or coupled cluster are necessary. Standard density functional methods (DFT) cannot capture these interactions due to the (semi-)local character of standard functionals. Only nonlocal functionals or functionals that explicitly contain a van der Waals term whose parameter is fitted to the experimental data can handle dispersive interactions. Applying the method of increments to such systems has not only the advantage that highly accurate CCSD(T) calculations are possible, but that we can also discuss the individual contributions to the adsorption energy and therefore gain better insight into the nature of bonding.

The ceria (110) and (111) surfaces are modeled with finite fragments, namely, $Ce_{12}O_{24}$ for the (110) surface and $Ce_{13}O_{26}$ for the (111) surface (see Figure 3.7). These fragments are embedded in a point charge array that is optimized so as to reproduce the electrostatic potential of the infinite system. For both surfaces, the molecules are adsorbed atop the cerium

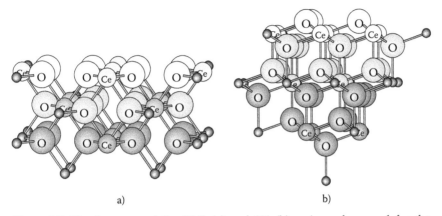

Figure 3.7 The fragment of the (110) (a) and 111 (b) ceria surface used for the correlation calculation. Point charges are not shown.

site. For CO, adsorption occurs in a perpendicular arrangement, with carbon pointing to the surface. For N_2O, it is under discussion whether it is adsorbed N-down or O-down and, in the latter case, whether it is in a perpendicular or tilted geometry. We tested all three possibilities in wavefunction-based correlated single-point calculations, relying on optimized DFT geometries. Results for adsorption energies, in comparison with other methods (DFT and HF), are listed in Table 3.6. We performed the incremental expansion both for an MP2 treatment of the correlation part and for a CCSD(T) treatment. The MP2 was used for comparing the incremental results with the full result, which is still feasible at the MP2 level but not at the CCSD(T) level. Overall, we achieved more than 95% of the correlation contribution to the adsorption energy with an incremental expansion involving two-body increments between the molecule and the surface ions up to the third-nearest neighbors and only the nearest-neighbor three-body increments. Comparing the different methods applied (we used the typical widely used density functional B3LYP), we can see a large difference in the performance for the chosen systems. Using the CCSD(T) values as our reference, we see that CO is 40% more strongly bound on the (111) surface than on the (110) one. The MP2 treatment yields about the same ratio, but the absolute values of the adsorption energy are 20% too large. B3LYP binds CO on both surfaces with about the same adsorption energy, and the absolute values are too small by more than a factor of 2. The Hartree–Fock treatment, finally, yields a stronger bond for the (110) than for the (111) surface, and the average of the adsorption energies is about the same as for B3LYP. Comparing the HF and CCSD(T) results, we can conclude that the correlation contribution to binding is much larger in the case of CO on (111) as compared to (110). Due to the different surface structure, the distance between the oxygen ions and the adsorbed molecule is much

Table 3.6 Adsorption Energies (meV) at Different
Levels of Theory

E^{ads}	B3LYP	HF	MP2	CCSD(T)
CO on 110	−104	−149	−244	−201
CO on 111	−110	−48	−323	−283
N_2O, N-down	−147	22	−276	−227
N_2O, O-down (perp)	6	4	−142	−133
N_2O, O-down (tilted)	−138	−105	−263	−261

Values reflect double-zeta basis sets for surface atoms and triple-zeta ones for the adsorbed molecules, for two different molecules on two different surfaces of ceria. The N_2O adsorption was only calculated for the (111) surface.

smaller for the (111) surface and the dispersive interaction is therefore stronger.

In the case of N_2O on ceria (111), binding is entirely due to correlation for the perpendicular adsorption, no matter whether it is N-down or O-down. Only in the case of the tilted O-down adsorption, Hartree–Fock yields some binding. At the B3LYP level, the N-down and O-down modes have almost the same interaction energy, with N-down slightly favored. However, the total interaction energy is much smaller (by a factor of 2) than was measured in the experiment (about −250 to −300 meV). At the MP2 level, the two adsorption geometries are still quite similar in energy, so that we cannot tell which is the more stable configuration. However, the total interaction energies are now in the range of the experimental values. Finally, at the highest level applied (i.e., the CCSD(T) level), the energetic order of the two adsorption modes changes, making the O-down mode the most stable adsorption mode by about 13%, which is consistent with the experiment. Individual correlation contributions to the adsorption energy of CO on ceria (110) are shown in Figure 3.8. Two surprising results have been found. First, about 40% of the correlation effect is due to the correlation of the filled semi-core $5sp$ shell of the Ce ion, on which the molecule adsorbs. Second, the correlation contributions of the surface and of the molecule are both drastically changed due to the adsorption. These parts are repulsive and cancel about three quarters of the attractive part. In order to test whether that is due to the dipole moment of CO or due to the ionic surface, we placed an argon atom at the same position as the CO molecule (see Figure 3.8). The results turned out to be similar; the repulsive parts are somewhat reduced in magnitude, but still amount to about 60% of the attractive part.

Another application of the method of increments to adsorption on surfaces concerns H_2S on a graphene layer [91]. Adsorption on graphene is expected to be dominated by dispersion forces, therefore highly accurate

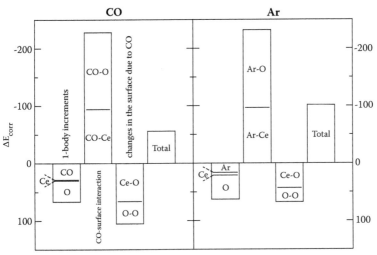

Figure 3.8 Individual correlation contributions (meV) to the adsorption energy of CO on the ceria (110) surface, in comparison to an absorbed argon atom at the same distance as the carbon atom.

quantum chemical correlation methods are necessary to obtain reliable results. We model the infinite graphene layer by finite fragments, saturating the dangling bonds with hydrogen atoms. The smallest possible fragment would be C_6, yielding a benzene-like molecule with the bond distances of graphene. The next larger fragment for an adsorption over the center of an C_6 ring would be C_{24}, an even larger fragment would be C_{54}. For the purpose of testing the convergence of the increments with the distance of the localized orbitals, we included all three clusters mentioned in our treatment. Therefore, we could only employ a quite small basis set of double-zeta quality at graphene and augmented double zeta with polarization functions at the molecule. For a detailed discussion of the basis sets effects see [91]. In Table 3.7, we list the individual contributions to the

Table 3.7 Individual Contributions (meV) to the Adsorption Energy of H_2S on a C_{24} Fragment of Graphene at the CCSD(T) Level

Hartree–Fock		+24.5
1-body-increments	(only cohesive part)	+16.3
2-body-increments	(surface-molecule)	−228.6
2-body-increments	(changes in the surface)	+2.7
3-body-increments		+8.2
Total correlation contribution to binding		−201.4
Total		−176.9

adsorption energy. As expected for a purely van der Waals bound system, the Hartree–Fock part is repulsive. Also, the contributions from the one-body increments, i.e., the changes in the correlation energy of the molecule and the surface compared to the free molecule or surface, respectively, are repulsive. The binding originates from the two-body increments between surface and molecule as expected for physisorbed molecules on ionic surfaces. In contrast to ionic surfaces, however, correlation energy changes within the surface are small for graphene and could even be neglected in determining the adsorption geometry. The three-body contributions are also small. To get a deeper insight as to where the binding originates, we separated the two-body surface-molecule contributions into two groups— those coming from the σ bonds and those from the π bonds. In Figure 3.9 we plot the individual contributions for all examined fragments. For the larger fragments, about 30% of the correlation contribution to adsorption comes from the σ bonds. They are not as long-range as the contributions from the π bonds, but are not negligible at all. Comparing the C_6 and the C_{54} fragments, where we have three localized π bonds at the inner ring in both cases, the contribution from C_6 has the same amount as that of the corresponding three localized orbital groups in C_{54}. These contributions account for about 60% of the whole π correlation effect on the adsorption energy.

In summary, the method of increments is well suited for investigating adsorption energies. Especially in the case of physisorbed molecules, it opens up the possibility of applying highly accurate quantum chemical correlation methods like coupled cluster to such systems. Another group has extended the method of increments to the adsorption of open-shell

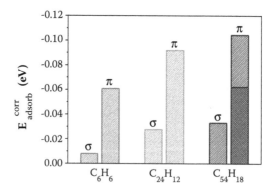

Figure 3.9 The two-body increments between surface and molecule for the adsorption of H_2S on graphene are plotted for different fragments and separated in σ-bond and π-bond contributions. The double shaded region in the π contribution of the C_{54} fragment indicates the increments of the inner C_6 ring.

atoms on ionic surfaces, namely Cu on a ZnO surface [92], applying an MRCEPA correlation treatment.

3.4 Conclusion and outlook

Since 1992, when the method of increments was first applied in the electronic theory of solids for determining correlation energies of diamond, silicon, and graphite, the method has been successfully applied to a wide variety of insulators and semiconductors, ranging from ionic solids over strongly bound covalent compounds to weakly bound van der Waals crystals.

The method has been extended to metallic systems in recent years. A number of problems arise here, the most prominent one concerning the localization of orbitals, and a second one concerning the modeling of these orbitals in finite fragments. To tackle these problems, we suggest basing the incremental expansion on well-localizable model systems. This way, the orbitals to be correlated are embedded in a surrounding that by itself has no metallic character but can mimic the metal in the interior region. Within the incremental scheme, we allow for a delocalization of the orbitals and therefore account for the metallic character with increasing order of the many-body expansion. Applications so far include group 2 and 12 metals where we obtained very good agreement with experiment. Further applications concern cases of metallic systems built up from open-shell atoms. For instance, the correlation energy of a metallic Li ring has been considered applying a projection technique for nonorthogonal occupied and virtual orbitals within the incremental scheme [40, 94]. Even a pilot study for 3-dimensional lithium has been successfully performed where the incremental scheme was based on closed-shell Li_2 subunits [95]. In this connection, a third problem occurring in metals has to be mentioned, namely the quasi-degeneracy of the orbitals around the Fermi surface. In such cases, a single-reference correlation method may fail, and the incremental scheme must then be based on high-order coupled-cluster [96] or multireference correlation methods [40]. This problem may arise not only in open-shell systems, but also, for example, in barium [97], where the d-bands cross the Fermi level.

In applying the method of increments to the adsorption energies of molecules on surfaces, we reformulated the method so that it applies to the correlation contributions to the adsorption energy. Two different applications, the adsorption of CO and N_2O on different ceria surfaces and H_2S on graphene, show that it is possible to apply highly accurate quantum chemical correlation methods to the adsorption problem. Another advantage of the method of increments is that one gains insight into the nature of bonding, because the individual contributions to the binding energy are available. For instance, one can discuss in detail the influence of the correlation of the semi-core shells, the distance behavior of the correlation effects, or the impact of three-body terms.

With the two extensions, to metals and adsorption processes, of the originally proposed method of increments at hand, it should be possible in the future to combine them, which would open the field of adsorption processes on metal surfaces.

Acknowledgments

We thank Prof. Dr. Peter Fulde (Dresden, Germany) for his guidance and a very fruitful collaboration over the last decades. The authors want to thank Dr. Elena Voloshina (Berlin, Germany), Dr. Nicola Gaston (Lower Hutt, New Zealand), Prof. Dr. Krzysztof Rosciscewski (Krakow, Poland), Dr. Ulrich Wedig (Stuttgart, Germany), Dr. Dirk Andrae (Berlin, Germany), and Dr. Carsten Müller (Uppsala, Sweden, now Berlin, Germany) for the intensive collaboration in various joint projects. Many valuable discussions with Prof. Dr. Peter Schwerdtfeger (Auckland, New Zealand), Prof. Dr. Kersti Hermannson (Uppsala, Sweden), and Prof. Dr. Martin Jansen (Stuttgart, Germany) are gratefully acknowledged by the authors. Financial support through the priority programs SPP 1145 (modern and universal first-principles methods for many-electron systems in chemistry and physics) and SPP 1178 (experimental electron density as key for understanding chemical interactions) of the DFG is acknowledged.

References

[1] P. Hohenberg and W. Kohn, Inhomogeneous electron gas, *Phys. Rev.* 136, B864 (1964).

[2] W. Kohn and L. J. Sham, Self-consistent equations including exchange and correlation effects, *Phys. Rev.* 140, A1133 (1965).

[3] R. M. Dreizler and E. K. U. Gross, *Density functional theory*, Springer, Berlin, 1990.

[4] H. Eschrig, *The fundamentals of density functional theory*, B. G. Teubner, Stuttgart, (1996).

[5] D. R. Hartree, The wave mechanics of an atom with a non-Coulomb central field. I. Theory and methods, *Proc. Cambridge Philos. Soc.* 24, 89 (1928).

[6] V. Fock, Näherungsmethode zur Lösung des quantenmechanischen Mehrkörperproblems, *Z. Phys.* 61, 126 (1930).

[7] J. C. Slater, Note on Hartree's method, *Phys. Rev.* 35, 210 (1930).

[8] P. O. Löwdin, Present situation of quantum chemistry, *J. Phys. Chem.* 61, 55 (1957).

[9] R. K. Nesbet, Approximate methods in the quantum theory of many-fermion systems, *Rev. Mod. Phys.* 33, 28 (1961).

[10] G. H. Wannier, The structure of electronic excitation levels in insulating crystals, *Phys. Rev.* 52, 191 (1937).

[11] G. Stollhoff and P. Fulde, A local approach to the computation of correlation energies of molecules, *Z. Phys. B* 26, 257 (1977).

[12] G. Stollhoff and P. Fulde, On the computation of electronic correlation energies within the local approach, *J. Chem. Phys.* 73, 4548 (1980).

[13] G. Stollhoff, The local ansatz extended, *J. Chem. Phys.* 105, 227 (1996).

[14] P. Pulay, Localizability of dynamic electron correlation, *Chem. Phys. Lett.* 100, 151 (1983).

[15] C. Pisani, M. Busso, G. Capecchi, S. Casassa, R. Dovesi, L. Maschio, C. Zicovich-Wilson and M. Schütz, Local-MP2 electron correlation method for nonconducting crystals, *J. Chem. Phys.* 122, 094113 (2005).

[16] C. Pisani, G. Capecchi, S. Casassa, L. Maschio, Computational aspects of a local MP2 treatment of electron correlation in periodic systems: SiC vs BeS, *Mol. Phys.* 03, 2527 (2005).

[17] S. Casassa, C. Zicovich-Wilson and C. Pisani, Symmetry-adapted localized Wannier functions suitable for periodic local correlation methods, *Theor. Chem. Acc.*, 116, 726 (2006).

[18] C. Pisani, S. Casassa and L. Maschio, On the prospective use of the one-electron density matrix as a test of the quality of post-Hartree–Fock schemes for crystals, *Z. Phys. Chem.* 220, 913 (2006).

[19] S. Casassa, M. Halo, L. Maschio, C. Roetti, C. Pisani, Beyond a Hartree–Fock description of crystalline solids: The case of lithium hydride, *Theor. Chem. Acc.* 117, 781 (2007).

[20] L. Maschio, D. Usvyat, C. Pisani, F. Manby, S. Casassa, M. Schütz, Fast local-MP2 method with density-fitting for crystals. I. Theory and algorithms, *Phys. Rev. B* 76, 075101 (2007).

[21] D. Usvyat, L. Maschio, F. Manby, M. Schütz, S. Casassa, C. Pisani, Fast local-MP2 method with density-fitting for crystals. II. Test calculations and application to the carbon dioxide crystal, *Phys. Rev. B* 76, 075102 (2007).

[22] C. Pisani, L. Maschio, S. Casassa, M. Halo, M. Schütz, D. Usvyat, Periodic local MP2 method for the study of electronic correlation in crystals: Theory and preliminary applications, *J. Comp. Chem.* 29, 2113 (2008).

[23] S. Casassa, M. Halo, L. Maschio, A local MP2 periodic study of crystalline argon, *J. Phys.: Conf. Ser.* 117, 011001 (2008).

[24] L. Maschio and D. Usvyat, Fitting of local densities in periodic systems, *Phys. Rev. B* 78, 073102 (2008).

[25] H. Stoll, Correlation energy of diamond, *Phys. Rev. B* 46, 6700 (1992).

[26] H. Stoll, The correlation energy of crystalline silicon, *Chem. Phys. Lett.* 191, 548 (1992).

[27] H. Stoll, On the correlation energy of graphite, *J. Chem. Phys.* 97, 8449 (1992).

[28] S. Kalvoda, B. Paulus, M. Dolg, H. Stoll, and H.-J. Werner, Electron correlation effects on structural and cohesive properties of closo-hydroborate dianions $(B_nH_n)^{2-}$ (n=5-12) and B_4H_4, *Phys. Chem. Chem. Phys.* 3, 514 (2001).

[29] H. Stoll, B. Paulus, and P. Fulde, On the accuracy of correlation-energy expansions in terms of local increments, *J. Chem. Phys.* 123, 144108 (2005).

[30] K. Doll, M. Dolg, P. Fulde, and H. Stoll, Correlation effects in ionic crystals: The cohesive energy of MgO, *Phys. Rev. B* 52, 4842 (1995).

[31] K. Doll, M. Dolg, and H. Stoll, Correlation effects in MgO and CaO: Cohesive energies and lattice constants, *Phys. Rev. B* 54, 13529 (1996).

[32] K. Doll and H. Stoll, Cohesive properties of alkali halides, *Phys. Rev. B* 56, 10121 (1997).

[33] K. Doll and H. Stoll, Ground-state properties of heavy alkali halides, *Phys. Rev. B* 57, 4327 (1998).

[34] K. Doll, P. Pyykkö, and H. Stoll, Closed-shell interaction in silver and gold chlorides, *J. Chem. Phys.* 109, 2339 (1998).

[35] B. Paulus, P. Fulde, and H. Stoll, Electron correlations for ground-state properties of group-IV semiconductors, *Phys. Rev. B* 51, 10572 (1995).

[36] B. Paulus, P. Fulde, and H. Stoll, Cohesive energies of cubic III-V semiconductors, *Phys. Rev. B* 54, 2556 (1996).

[37] S. Kalvoda, B. Paulus, P. Fulde, and H. Stoll, Influence of electron correlations on ground-state properties of III-V semiconductors, *Phys. Rev. B* 55, 4027 (1997).

[38] B. Paulus, F.-J. Shi, and H. Stoll, A correlated *ab initio* treatment of the zinc-blende wurtzite polytypism of SiC and III - V nitrides, *J. Phys.: Condens. Matter* 9, 2745 (1997).

[39] M. Albrecht, B. Paulus, and H. Stoll, Correlated ab initio calculations for ground-state properties of II-VI semiconductors, *Phys. Rev. B* 56, 7339 (1997).

[40] B. Paulus, Towards an *ab initio* incremental correlation treatment of metals, *Chem. Phys. Lett.* 371, 7 (2003).

[41] B. Paulus and K. Rościszewski, Metallic bonding due to electronic correlations: a quantum chemical *ab initio* calculation of the cohesive energy of mercury, *Chem. Phys. Lett.* 394, 96 (2004).

[42] B. Paulus, K. Rościscewski, N. Gaston, P. Schwerdtfeger, and H. Stoll, The convergence of the *ab initio* many-body expansion for the cohesive energy of solid mercury, *Phys. Rev. B* 70, 165106 (2004).

[43] N. Gaston, B. Paulus, K. Rościscewski, P. Schwerdtfeger, and H. Stoll, Lattice structure of mercury: Influence of electronic correlation, *Phys. Rev. B* 74, 094102 (2006).

[44] E. Voloshina and B. Paulus, Correlation energies for small magnesium clusters in comparison with bulk magnesium, *Mol. Phys.* 105, 2849 (2007).

[45] E. Voloshina and B. Paulus, Wavefunction-based *ab initio* method for metals: application of the incremntal scheme to magnesium, *Phys. Rev. B.* 75, 245117 (2007).

[46] E. Voloshina, N. Gaston, and B. Paulus, Embedding procedure for *ab initio* correlation calculations in group II metals, *J. Chem. Phys.* 126, 134115 (2007).

[47] N. Gaston and B. Paulus, *Ab initio* calculations for the ground-state properties of group-12 metals Zn and Cd, *Phys. Rev. B* 76, 214116 (2007).

[48] E. Voloshina, B. Paulus, and H. Stoll, Quantum-chemical approach to cohesive properties of metallic beryllium, *J. Phys. Conf. Ser.* 117, 012029 (2008).

[49] N. Gaston, B. Paulus, U. Wedig, and M. Jansen, Multiple minima on the energy landscape of elemental zinc: a wave function based ab initio study, *Phys. Rev. Lett.* 100, 226404 (2008).

[50] N. Gaston, D. Andrae, B. Paulus, U. Wedig, and M. Jansen, Understanding the hcp anisotropy in Cd and Zn: the role of electron correlation in determining the potential energy surface, *Phys. Chem. Chem. Phys.* 12, 618 (2010).

[51] K. Rościszewski, B. Paulus, P. Fulde, and H. Stoll, *Ab initio* calculation of ground-state properties of rare-gas crystals, *Phys. Rev. B* 60, 7905 (1999).

[52] K. Rościszewski, B. Paulus, P. Fulde, and H. Stoll, *Ab initio* coupled-cluster calculations for the fcc and hcp structure of rare-gas solids, *Phys. Rev. B* 62, 5482 (2000).

[53] K. Rościszewski and B. Paulus, Influence of three-body forces and anharmonic effects on the zero-point energy of rare-gas crystals, *Phys. Rev. B* 66, 092102 (2002).

[54] B. Paulus, Wave-function-based *ab initio* correlation treatment for the buckminsterfullerene C60, *Int. J. Quant. Chem.* 100, 1026 (2004).

[55] M. Yu, S. Kalvoda, and M. Dolg, An incremental approach for correlation contributions to the structural and cohesive properties of polymers. Coupled-cluster study of trans-polyacetylene, *Chem. Phys.* 224, 121 (1997).

[56] A. Abdurahman, M. Albrecht, A. Shukla, and M. Dolg, *Ab initio* study of structural and cohesive properties of polymers: Polyiminoborane and polyaminoborane, *J. Chem. Phys.* 110, 8819 (1999).

[57] A. Abdurahman, A. Shukla, and M. Dolg, *Ab initio* treatment of electron correlations in polymers: Lithium hydride chain and beryllium hydride polymer, *J. Chem. Phys.* 112, 4801 (2000).

[58] A. Abdurahman, A. Shukla, and M. Dolg, Correlated ground-state *ab initio* calculations of polymethineimine, *Chem. Phys.* 257, 301 (2000).

[59] A. Abdurahman, A. Shukla, and M. Dolg, *Ab initio* many-body calculations on infinite carbon and boron-nitrogen chains, *Phys. Rev. B* 65, 115106 (2002).

[60] C. Willnauer and U. Birkenheuer, Quantum chemical *ab initio* calculations of correlation effects in complex polymers: Poly(para-phenylene), *J. Chem. Phys.* 120, 11910 (2004).

[61] C. Buth and B. Paulus, Basis set convergence in extended systems: infinite hydrogen fluoride and hydrogen chloride chains, *Chem. Phys. Lett.*, 398, 44 (2004).

[62] C. Buth and B. Paulus, Hydrogen bonding in infinite hydrogen fluoride and hydrogen chloride chains, *Phys. Rev. B,* 74 045122 (2006).

[63] K. Doll, M. Dolg, P. Fulde, and H. Stoll, Quantum chemical approach to cohesive properties of NiO, *Phys. Rev. B* 55, 10282 (1997).

[64] K. Rościszewski, K. Doll, B. Paulus, P. Fulde, and H. Stoll, Ground-state properties of rutile: Electron-correlation effects, *Phys. Rev. B* 57, 14667 (1998).

[65] S. Kalvoda, M. Dolg, H.-J. Flad, P. Fulde, and H. Stoll, *Ab initio* approach to cohesive properties of GdN, *Phys. Rev. B* 57, 2127 (1998).

[66] E. Voloshina and B. Paulus, On the application of the incremental scheme to ionic solids: test of different embeddings, *Theor. Chem. Acc.* 114, 259 (2005).

[67] E. Voloshina and B. Paulus, Influence of electronic correlations on the ground-state properties of cerium dioxide, *J. Chem. Phys.* 124, 234711 (2006).

[68] E. Voloshina and B. Paulus, Cohesive properties of CeN and LaN from first principles, *J. Comput. Chem.* 29, 2107 (2008).

[69] B. Paulus, The method of increments, a wavefunction-based *ab initio* correlation method for solids, *Phys. Rep.* 428, 1 (2006).

[70] E. Voloshina and B. Paulus, Wavefunction-based *ab initio* correlation methods for metals: Application of the incremental scheme to Be, Mg, Zn, Cd, and Hg, in *Chemical modelling: Applications and theory*, Vol. 6, M. Springborg (ed.) R.S.C, Cambridge, 2009.

[71] R. Dovesi, V. R. Saunders, C. Roetti, R. Orlando, C. M. Zicovich-Wilson, F. Pascale, B. Civalleri, K. Doll, N. M. Harrison, I. J. Bush, Ph. DArco, M. Lunell, CRYSTAL06, User Manual (Torino, University of Torino, 2006).

[72] J. M. Foster and S. F. Boys, Construction of some molecular orbitals to be approximately invariant for changes from one molecule to another, *Rev. Mod. Phys.* 32, 296 (1960).

[73] J. Pipek and P. G. Mezey, A fast intrinsic localization procedure applicable for *ab initio* and semiempirical linear combination of atomic orbital wave functions, *J. Chem. Phys.* 90, 4916 (1989).

[74] R. K. Nesbet, Atomic Bethe-Goldstone equations, *Adv. Chem. Phys.* 14, 1 (1969).

[75] P. Fulde, Wavefunction methods in electronic-structure theory of solids, *Adv. in Physics* 51, 909 (2002).

[76] R. D. Amos, A. Bernhardsson, A. Berning, P. Celani, D. L. Cooper, M. J. O. Deegan, A. J. Dobbyn, F. Eckert, C. Hampel, G. Hetzer, P. J. Knowles, T. Korona, R. Lindh, A. W. Lloyd, S. J. McNicholas, F. R. Manby, W. Meyer, M. E. Mura, A. Nicklass, P. Palmieri, R. Pitzer, G. Rauhut, M. Schütz, U. Schumann, H. Stoll, A. J. Stone, R. Tarroni, T. Thorsteinsson, and H.-J. Werner, MOLPRO, a package of ab initio programs designed by H.-J. Werner and P. J. Knowles, Birmingham, 2006.

[77] C. Hampel, K. Peterson, and H.-J. Werner, A comparison of the efficiency and accuracy of the quadratic configuration interaction (QCISD), coupled cluster (CCSD), and Brueckner coupled cluster (BCCD) methods, *Chem. Phys. Lett.* 190, 1 (1992).

[78] M. J. O. Deegan and P. J. Knowles, Perturbative corrections to account for triple excitations in closed and open shell coupled cluster theories, *Chem. Phys. Lett.* 227, 321 (1994).

[79] J. Friedrich, M. Hanrath, and M. Dolg, Fully automated implementation of the incremental scheme: Application to CCSD energies for hydrocarbons and transition metal compounds, *J. Chem. Phys.* 126, 154110 (2007).

[80] J. Friedrich, M. Hanrath, and M. Dolg, Error analysis of incremental electron correlation calculations and applications to clusters and potential energy surfaces, *Chem. Phys.* 338, 33 (2007).

[81] J. Friedrich, M. Hanrath, and M. Dolg, Energy Screening for the Incremental Scheme: Application to Intermolecular Interactions, *J. Phys. Chem. A* 111, 9830 (2007).

[82] J. Friedrich and M. Dolg, Implementation and performance of a domain-specific basis set incremental approach for correlation energies: Applications to hydrocarbons and a glycine oligomer, *J. Chem. Phys.* 129, 244105 (2008).

[83] J. Friedrich, M. Hanrath, and M. Dolg, Using symmetry in the framework of the incremental scheme: Molecular applications, *Chem. Phys.* 346, 266 (2008).

[84] J. Friedrich, K. Walczak, and M. Dolg, Evaluation of core and core, valence correlation contributions using the incremental scheme, *Chem. Phys.* 356, 47 (2009).

[85] J. Friedrich, M. Hanrath, and M. Dolg, Evaluation of incremental correlation energies for open-shell systems: Application to the intermediates of the 4-exo cyclization, arduengo carbenes and an anionic water cluster, *J. Phys. Chem. A* 112, 8762 (2008).

[86] J. Friedrich and M. Dolg, Fully automated incremental evaluation of MP2 and CCSD(T) energies: Application to water clusters, *J. Chem. Theo. Comp.* 5, 287 (2009).

[87] S. F. Boys and F. Bernardi, The calculation of small molecular interactions by the differences of separate total energies. Some procedures with reduced errors, *Mol. Phys.* 19, 553 (1970).

[88] C. Müller, B. Herschend, K. Hermansson, and B. Paulus, Application of the method of increments to the adsorption of CO on the $CeO_2(110)$ surface, *J. Chem. Phys.* 128, 214701 (2008).

[89] C. Müller, K. Hermansson, and B. Paulus, Electron correlation contribution to the N_2O/ceria(111) interaction, *Chem. Phys.* 362 (2009) 91.

[90] C. Müller, B. Paulus, and K. Hermansson, *Ab initio* calculations of CO physisorption on ceria(111), *Surf. Sci.* 603, 2619 (2009).

[91] B. Paulus and K. Rosciszewski, Application of the method of increments to the adsorption of H_2S on graphene, *Int. J. Quantum Chem.* 109, 3055 (2009).

[92] I. Schmitt, K. Fink, and V. Staemmler, The method of local increments for the calculation of adsorption energies of atoms and small molecules on solid surfaces. Part I. A single Cu atom on the polar surfaces of ZnO. *Phys. Chem. Chem. Phys.* 11, 11196 (2009).

[93] U. Wedig, M. Jansen, B. Paulus, K. Rosciszewski, and P. Sony, Structural and Electronic Properties of Mg, Zn and Cd, *Phys. Rev. B* 75, 205123 (2007).

[94] H. Stoll, Towards an Incremental Expansion of Strong Correlation Effects in Solids, *Ann. Phys.*, 5 (1996) 355.

[95] H. Stoll, B. Paulus, and P. Fulde, An incremental coupled-cluster approach to metallic lithium, *Chem. Phys. Lett.* 469 (2009) 90.

[96] H. Stoll, Toward a wave-function-based treatment of metals: extrapolation from finite clusters, *J. Phys. Chem. A* 113, 11483 (2009).

[97] B. Paulus and A. Mitin, An incremental method for the calculation of the electron correlation energy in metals, *Lecture Series on Computer and Computational Science* 7, 935 (2006).

chapter four

The hierarchical scheme
for electron correlation
in crystalline solids

Stephen J. Nolan, Peter J. Bygrave, Neil L. Allan, Michael J. Gillan, Simon J. Binnie, and Frederick R. Manby

Contents

4.1 Introduction

There appear to be three main ways to go beyond a density-functional-theory (DFT) description of the solid state:

1. Quantum Monte Carlo methods
2. Periodic wavefunction-based electronic structure theory
3. Extraction of information about the bulk from high-level calculations on finite clusters

The first is well established, especially in the physics community, and enjoys the distinctions of being exact in principle and exceedingly parallelizable [1]. The well-known drawbacks arise from intrinsic errors if the fixed-node approximation is used (but see, for example, [2] and [3]), the computational cost of reducing variance in the energies, and the associated challenges of computing accurate gradients (but see, for example, [4] and [5]).

The advantage of periodic wavefunction-based electronic structure theory (described, for example, in Chapters 1 and 2, and in the references they cite) derives from a direct connection to the successful hierarchies of molecular electronic structure theory, but this comes at the cost of exceedingly complicated implementation.

Extraction of information from calculations on clusters is certainly very simple, but is open to the criticism that calculations are restricted to relatively small fragments, where the surface, not the bulk-like interior, dominates the properties of the cluster. One means of remedying this problem is to embed clusters in some simplified representation of the bulk.[1] Instead we discuss here a simple approach for accurately eliminating surface effects through combination of results from clusters of various sizes.

We begin by considering the cohesive energy of an ionic solid MX in the rock salt structure. Here, cohesive energies will be expressed per formula unit and relative to free atoms. The static cohesive energy consists of a Hartree–Fock (HF) term and a smaller correlation contribution:

$$E_{coh}^{static} = E_{coh}^{HF} + E_{coh}^{corr}. \tag{4.1}$$

The process of dissociating the crystal into atoms can be divided into the separation of the crystal into MX molecules with the bulk interatomic distance, followed by dissociation of the MX molecule into its constituent atoms. For lithium hydride we found that the correlation contribution E_{coh}^{corr} to the static cohesive energy was dominated by the correlation contribution to the dissociation energy of the LiH diatomic molecule (with an internuclear separation equal to that in the bulk solid) into atoms [6]. We therefore express the correlation contribution to the cohesive energy as a sum of two terms giving

$$E_{coh}^{static} = E_{coh}^{HF} + E_{mol}^{corr} + \Delta E_{coh}^{corr}. \tag{4.2}$$

The various energy contributions are illustrated in Figure 4.1.

Assuming that the Hartree–Fock contribution E_{coh}^{HF} can be evaluated accurately using periodic Hartree–Fock theory,[2] and the molecular term E_{mol}^{corr} using conventional molecular electronic structure theory, we are left with the problem of computing what we call the *correlation residual*, ΔE_{coh}^{corr}.

To understand how best to approximate the correlation residual, we begin by considering the simple case of a one-dimensional array of MX atoms. Let the correlation energy relative to separated MX molecules of the chain of $2n$ atoms (or n MX units) be ΔE_{2n}^{corr}. The correlation contribution to the binding energy of the chain per MX unit could be evaluated through the limit

$$\Delta E_{bind}^{corr} = \lim_{n \to \infty} \frac{\Delta E_{2n}^{corr}}{n}. \tag{4.3}$$

---

[1] Embedded electronic structure theory for condensed-phase problems constitutes a major field in its own right, which is beyond the scope of the present volume.

[2] Although this turns out not to be particularly straightforward — see Ref. 7 and Section 4.3.

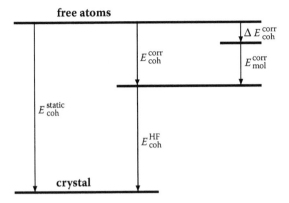

Figure 4.1 Illustration of the various contributions to the static cohesive energy E_{coh}^{static} of an ionic crystal. The energy is first divided into Hartree–Fock (E_{coh}^{HF}) and correlation (E_{coh}^{corr}) contributions; the correlation contribution is dominated by the dissociation of a *molecule* MX into its constituent atoms (E_{mol}^{coh}) with a small correction ΔE_{coh}^{corr} for the dissociation of the crystal into molecular fragments.

In practice, this procedure converges very slowly, because until n is very large the effects of the ends of the chain make a significant contribution to the energetics.

A more robust scheme is to look at the energy change on adding one formula unit to a chain of length $2n$ to form one of length $2n + 2$:

$$\Delta E_{bind}^{corr} = \lim_{n \to \infty} \left(\Delta E_{2n+2}^{corr} - \Delta E_{2n}^{corr} \right). \tag{4.4}$$

In this expression, the effects of the ends of the chain largely cancel out, and convergence to the limit is much more rapid than in Equation (4.3). This is illustrated in Figure 4.2 for lithium hydride, and was demonstrated in much earlier work by Abdurahman et al. [8].

The hierarchical method can be seen as the 3-dimensional generalization of this technique [6]. Consider a cluster containing an integer number of MX ions with positions taken from the rock salt structure (and with no relaxation of the geometry of the cluster). Let the correlation energy residual of the $\ell \times m \times n$ cluster be $\Delta E_{\ell mn}^{corr} \equiv E_{\ell mn}$. In what follows, we use the simplified notation $E_{\ell mn}$ for clarity, and refer to these quantities as energies. If two side-lengths are fixed, say ℓ and m, then the energy will vary linearly with respect to n, provided n is large enough:

$$E_{\ell mn} = E_{\ell m}^0 + nE_{\ell m}^1. \tag{4.5}$$

Here the slope of $E_{\ell mn}$ with respect to n is given by the coefficient $E_{\ell m}^1$, and the intercept $E_{\ell m}^0$ arises because of the edge effects. The coefficients $E_{\ell m}^\nu$

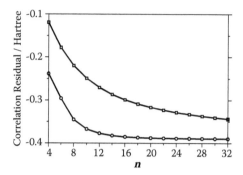

Figure 4.2 Convergence of the correlation residual $\Delta E_{\text{bind}}^{\text{corr}}$ for a linear chain of lithium hydride with respect to the number of atoms in the chain, computed by the simple ratio $\Delta E_{2n}^{\text{corr}}/n$ (squares) and by chain extension ($\Delta E_{2n+2}^{\text{corr}} - \Delta E_{2n}^{\text{corr}}$) (circles). The correlation energies were obtained using the MP2/cc-pVTZ level of theory.

themselves obey similar relationships, so that for fixed ℓ we have

$$E_{\ell m}^{\nu} = E_{\ell}^{0\nu} + m E_{\ell}^{1\nu} \tag{4.6}$$

and similarly

$$E_{\ell}^{\mu\nu} = E^{0\mu\nu} + \ell E^{1\mu\nu}. \tag{4.7}$$

Substituting Equation (4.7) into Equation (4.6), substituting the result into Equation (4.5), and using symmetry to simplify the expression we find

$$E_{\ell mn} = E^{000} + (\ell + m + n)E^{001} + (\ell m + \ell n + mn)E^{011} + \ell mn E^{111}. \tag{4.8}$$

This equation—the so-called *hierarchical equation*—expresses the energy of a finite cluster in terms of contributions, which scale as L^0, L^1, L^2, and L^3, respectively, where L is a measure of extension in one dimension. These contributions correspond to energies of corners, edges, surfaces, and the bulk. The coefficients $E^{\lambda\mu\nu}$ can be obtained by choosing four distinct clusters and solving a set of simultaneous equations; the coefficient E^{111} corresponds to the correlation residual contribution to the static bulk cohesive energy: $E^{111} = \Delta E_{\text{coh}}^{\text{corr}}$. As shown later, this estimate of the correlation residual converges very quickly with cluster size.

A different, and in some ways more physically transparent derivation is possible. We assume that the contribution of an atom depends only on its coordination number and thus, for a cuboid cluster, we can write

$$E_{\ell+2,m+2,n+2} = 8\tilde{E}^{000} + 4(\ell + m + n)\tilde{E}^{001} + 2(\ell m + \ell n + mn)\tilde{E}^{011} + \ell mn\tilde{E}^{111} \tag{4.9}$$

based on the number of atoms found at the corners, the edges, the surfaces, and in the interior, respectively. This more explicitly reveals the coefficient $\tilde{E}^{111}$ as the relevant contribution to the static bulk cohesive energy, and in fact it can be shown that $\tilde{E}^{111} \equiv E^{111}$. Moreover, the other coefficients now have physical meaning; for example, the $\tilde{E}^{011}$ coefficient relates to the surface formation energy (see Section 4.2.2).

To determine the four coefficients in Equation (4.8) or (4.9), it is necessary to choose four clusters. For maximum accuracy these should be chosen to be as bulk-like as possible; in other words, if a maximum of N ions can be used, the worst choice would be the chain $1 \times 1 \times N$. In fact, the bulk-like clusters appear naturally at the end of a lexicographically ordered list of clusters up to a given maximum size N, and we typically select the last four clusters in such a list to solve the simultaneous equations that produce bulk properties. For some values of N the equations are not soluble for this choice (for example, if all four share a common value of ℓ, m, or n), so the full prescription is to select the four clusters closest to the end of the list for which the equations have a solution.

4.2 Overview of results

4.2.1 Properties of crystalline lithium hydride

Earlier theoretical studies on lithium hydride have applied DFT (for example, see [9]) or the method of increments [10] (see also Chapter 3 of this book) with some success; however, neither method has been able to predict the properties of the crystal with accuracy close to that of the experiment. There are well-known deficiencies in DFT, such as the lack of dispersion, self-interaction errors, and a lack of systematic improvability, and results typically depend strongly on the functional used [11].

The hierarchical method provides a simple means by which the systematically improvable techniques of quantum chemistry can be applied to compute the correlation component of the cohesive energy, and these can then be combined with highly converged periodic Hartree–Fock energies to obtain the total cohesive energy at any volume. We now summarize how this combination accurately reproduces experimentally observed properties of lithium hydride and deuteride [12].

The Hartree–Fock (HF) approximation to the cohesive energy of a crystal can be calculated in the infinite basis set limit using plane waves provided pseudopotentials are used. In order to achieve the desired accuracy, the pseudopotential error must then be carefully subtracted. A suitable correction can be calculated using the hierarchical method, by calculating the Hartree–Fock energy for a series of clusters using both pseudo-potential and all-electron methods. The pseudo-potential error in the cohesive energy of the bulk system can then be estimated through use

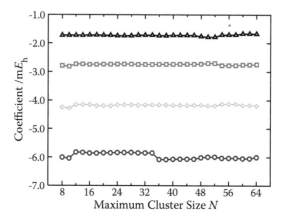

Figure 4.3 Coefficients from the hierarchical equation for LiH, $a = 4.084$ Å, as a function of maximum cluster size N calculated using MP2/cc-pVTZ. The parameter $\tilde{E}^{000}$ is shown as triangles, $\tilde{E}^{001}$ as squares, $\tilde{E}^{011}$ as diamonds, and $\tilde{E}^{111} \equiv E^{111}$ as circles.

of the hierarchical equation [7]. This procedure allows the Hartree–Fock energy of lithium hydride to be converged to an accuracy that was at the time unprecedented ($\pm 1 \, \mu E_h$). It also became apparent that previous Hartree–Fock calculations (e.g., by Casassa et al. [13]) on lithium hydride are not fully converged with respect to basis set size. A fuller discussion comparing the results of this study with other, more recent work is given in Section 4.3.

The molecular contribution to the cohesive energy, E^{corr}_{mol}, can be calculated to very high accuracy using existing quantum chemistry techniques and will not be discussed in detail here. There are, however, a number of potential sources of error in the correlation residual that must be overcome if an overall accuracy of around one tenth or a few tenths of a millihartree in the cohesive energy is to be achieved.

All the molecular electronic structure calculations here were performed using the MOLPRO package [14]. Using the hierarchical algorithm outlined in the introduction, the hierarchical equation was solved simultaneously for the coefficients $\tilde{E}^{000}$, $\tilde{E}^{001}$, $\tilde{E}^{011}$, and $\tilde{E}^{111}$. Some typical results for lithium hydride are shown in Figure 4.3 as a function of maximum cluster size N. The values of the coefficients do not change wildly with cluster size: the assumption that electron correlation effects on binding are short ranged therefore seems to be valid. Nevertheless, while there is little variation with cluster size on this scale, this is not sufficient to achieve our desired accuracy. Only the bulk coefficient E^{111} contributes to the cohesive energy, and while we concentrate on this quantity, the conclusions are general to all of the hierarchical solutions. Results for E^{111} from Figure 4.3 are shown in Figure 4.4 on a larger scale.

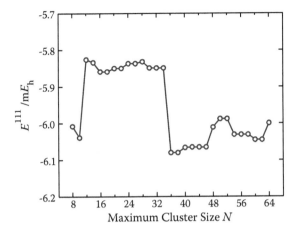

Figure 4.4 E^{111} in mE_h plotted as a function of maximum cluster size N for LiH, $a = 4.084$ Å, using MP2/cc-pVTZ.

In Figure 4.4 two step-changes are clearly visible at $N = 12$ and $N = 36$. For very small values of N, chains ($1 \times 1 \times n$) and sheets ($1 \times m \times n$) are required to solve the hierarchical equation. As cluster size increases, there comes a point at which chains no longer need to be included in order to solve the hierarchical equation, and for lithium hydride this corresponds to $N = 12$. Similarly, once the maximum cluster size has reached 36, sheets are no longer needed. The value $N = 36$ also marks the first point at which a cluster with atoms at bulk-like (i.e., fully enclosed) sites is included ($3 \times 3 \times 4$). From this point on, all clusters considered have at least two atoms along each axis ($l, m, n \geq 2$). These clusters clearly represent the bulk crystal more accurately than sheets or chains of atoms. That very small cluster sizes ($N < 12$) give similar values for E^{111} is coincidental; this is not observed for either lithium fluoride or neon.

Once clusters of at least 36 atoms have been reached, there is very little change in the value of E^{111}, and the remaining variation is much smaller than our target accuracy of around 0.1 mE_h.

It is reassuring to establish that the algorithm employed to select clusters as a function of maximum cluster size does not produce misleading results. As noted above, the algorithm selects one set of four clusters for a given maximum cluster size, which allows the hierarchical equation to be solved, producing one value for E^{111}. By considering a number of different sets of clusters, a range of values for E^{111} can be produced and compared to the original hierarchical result. For a given maximum cluster size N, all possible sets were generated from a range of clusters with no more than N atoms and no fewer than the smallest cluster from the original $N-2$ set. The

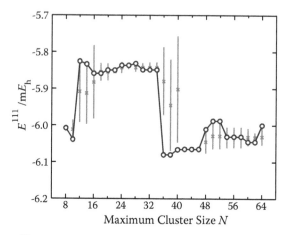

Figure 4.5 The E^{111} coefficients from the original hierarchical result (circles) and the average of all possible cluster combinations (crosses) at particular maximum cluster size N. The vertical lines indicate one standard deviation from the average. Both results are based on the same cluster energies from MP2/cc-pVTZ calculations.

average of these results as a function of maximum cluster size is compared with the original hierarchical result in Figure 4.5 (cf. [12]).

The step-changes observed earlier in Figure 4.4 are still present in Figure 4.5 but are more gradual. The large standard deviations at these points are consistent with our understanding of the original hierarchical result. Around the first step-change, some sets will still contain chains and give a low value of E^{111} while others without chains lead to a high E^{111}, resulting in a large standard deviation. Once chains are eliminated from all clusters, the average values are close to those in Figure 4.4 and the standard deviation decreases significantly. Similar behavior is observed at the second step-change, after which the average remains almost constant and the original hierarchical results are representative of this average. The step-changes will occur regardless of the specific algorithm that is used; the current algorithm is simply more efficient in passing through the second step-change at smaller cluster sizes than an averaged method.

Calculations at rather small cluster sizes can be used to give good approximations for small energy corrections based on higher levels of theory and larger basis sets. To demonstrate this, a range of calculations have been performed for the same lithium hydride clusters (see Figure 4.6). Each set of results in Figure 4.6 has the now-familiar hierarchical shape and is roughly parallel to one another. This suggests that the difference between results from different methods and basis sets converges more rapidly with cluster size than the absolute value of E^{111}, and this can be seen clearly in

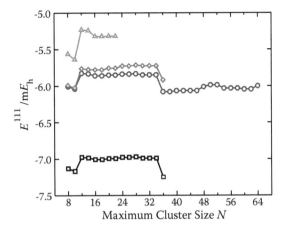

Figure 4.6 E^{111} coefficient results for LiH, a = 4.084 Å, as a function of maximum cluster size N for different methods and basis sets: MP2/cc-pVTZ in circles, MP2/cc-pVQZ in diamonds, ae-MP2/cc-pCVTZ in squares, and CCSD(T)/cc-pVTZ in triangles.

Figure 4.7, in which the same data are plotted relative to the MP2/cc-pVTZ result.

The trends shown in Figure 4.7 allow corrections to be made at small cluster sizes without introducing large errors. Without performing these corrections at small cluster sizes, it would not be possible to include these levels of theory in the calculation of E^{111}. The error introduced by calculating these corrections at small cluster sizes is significantly smaller than

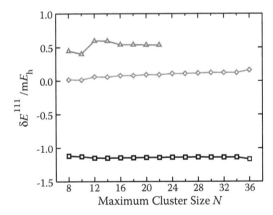

Figure 4.7 E^{111} coefficient results for LiH, $a = 4.084$ Å, plotted as a difference from MP2/cc-pVTZ: MP2/cc-pVQZ in diamonds, ae-MP2/cc-pCVTZ in squares, and CCSD(T)/cc-pVTZ in triangles.

Table 4.1 Contributions to the
Cohesive Energy of Lithium Hydride
at $a = 4.084$ Å in Millihartree

Contribution	Energy/mE_h
$E_{\text{static}}^{\text{HF}}$	−131.90
$E_{\text{mol}}^{\text{corr}}$	−38.40
$2E^{111}$ MP2/cc-pVTZ	−12.00
δcore	−2.29
δCCSD(T)	+1.08
δbasis	+0.80
ZPE PBE phonons	+7.79
$E_{\text{coh}}^{0\,\text{K}}$	−174.92

not including them at all: the errors associated with performing these corrections at $N = 16$ rather than at $N = 36$ are less than ± 0.1 mE_h. There is a clear upward trend in the basis set result, which indicates that a further correction is needed for an even larger basis set. For a more detailed discussion of corrections, including contributions from higher orders of coupled cluster theory and the diagonal Born–Oppenheimer correction, see [12].

In summary, the algorithm used to select the clusters does not distort the results for E^{111} and in order to ensure accurate results a reference calculation must be made at a cluster size of at least $N = 36$. Corrections for basis set and level of theory should of course be made using the largest clusters possible, but it appears that the accuracy of the reference calculation is maintained if these corrections are computed with $N = 16$. Typical results from this procedure for lithium hydride are shown in Table 4.1.

From Table 4.1, the importance of zero-point energy effects for lithium hydride is clear. As these calculations are not connected with the hierarchical method itself, they will not be discussed here. By repeating the procedure outlined in Table 4.1 at a range of different lattice parameters, an equation of state can be determined for the system and key properties derived. Table 4.2 shows the exceptionally good agreement between the hierarchical and experimental results for lithium hydride and deuteride.

4.2.2   Surface (001) energy of LiH

The coefficients in the hierarchical equation can also be used to calculate the surface formation energies of a crystal. As these have already been calculated in determining the bulk cohesive energy, this requires no further computational effort. The surface formation energy can be separated into two terms in a similar way to the cohesive energy

$$\sigma_{\text{static}} = \sigma^{\text{HF}} + \sigma^{\text{corr}}. \tag{4.10}$$

Table 4.2 Comparison of Experimental and Theoretical Lattice Parameters, Bulk Moduli and Cohesive Energies for Lithium Hydride and Lithium Deuteride

	Property	Experiment	Hierarchical
LiH	$a/\text{Å}$	4.061	4.062
	B_0/GPa	33–38	33.2
	$E_{\text{coh}}^{0\,K}/mE_h$	−174.9	−175.3
LiD	$a/\text{Å}$	4.045	4.046
	B_0/GPa	33.5	34.2
	$E_{\text{coh}}^{0\,K}/mE_h$	−177.5	−177.0

Source: S. J. Nolan, M. J. Gillan, D. Alfè, N. L. Allan, and F. R. Manby, *Phys. Rev. B* 80, 165109 (2009).

A description of how the Hartree–Fock component is calculated is given in [15]. As the molecular component of the correlation contribution will remain unchanged, only hierarchical terms need be considered. To obtain the correlation contribution σ^{corr} to the surface energy, the change in energy must be scaled by the surface area created, so for the (001) surface of a rock-salt structure

$$\sigma^{\text{corr}} = \frac{\tilde{E}^{011} - \tilde{E}^{111}}{a^2/4}. \tag{4.11}$$

This has been computed for lithium hydride at the experimentally determined lattice parameter $a = 4.084$ Å, using the same level of theory as was used for the cohesive energy (see Table 4.1). The surface energy has also been studied using a variety of density functionals, producing a range of values, as well as quantum Monte Carlo techniques [11]. The results are summarized in Table 4.3. The hierarchical method is in good agreement with the quantum Monte Carlo result. There is as yet no experimental value available for comparison, but if one becomes available (or in the case of other systems) accurate agreement will require consideration of the effects of vibrational terms and surface relaxation, neither of which is expected to be particularly large for this surface [11].

4.2.3 Lithium fluoride

Preliminary calculations on lithium fluoride have also been performed in order to demonstrate the suitability of the hierarchical scheme for larger systems. In principle it should be possible to achieve a similar level of accuracy for lithium fluoride as for lithium hydride. The experimental cohesive

Table 4.3 Contributions to the Surface Formation Energy of LiH a = 4.084 Å

Method	E^{011}/mE_h	E^{111}/mE_h	$\sigma/\mathrm{J\,m^{-2}}$
HF[a]			0.1893
Hierarchy[b]	−4.196	−6.000	+0.1886
δcore[c]	−0.840	−1.147	+0.0321
δCCSD(T)[d]	+0.627	+0.540	+0.0091
δbasis[e]	+0.277	+0.280	−0.0003
$\sigma_{\text{frozen core}}$			0.3867
σ_{total}			0.4188
DFT			0.2723 − 0.4663
DMC			0.36(1)

Note: The σ_{total} value is the hierarchical result computed as the sum of the Hartree–
     Fock and correlation contributions above it; $\sigma_{\text{frozen core}}$ excludes the core correction
     and is shown for comparison with the frozen-core diffusion Monte Carlo (DMC)
     result [11]. The DFT results were obtained with a variety of functionals from S. J.
     Binnie, E. Sola, D. Alfè, and M. J. Gillan, *Molecular Simulation* 35, 609 (2009).
[a] M. Marsman, A. Grüneis, J. Paier, and G. Kresse, *J. Chem. Phys.* 130, 184103 (2009).
[b] Based on MP2/cc-pVTZ $N = 64$
[c] Based on MP2/cc-pCVTZ $N = 16$
[d] Based on CCSD(T)/cc-pVTZ $N = 16$
[e] Based on MP2/cc-pV[TQ]Z $N = 16$

energy of lithium fluoride is 323.5 mE_h [16]. The close agreement with the
results shown in Table 4.4 is somewhat fortuitous. These calculations were
performed at the room temperature experimental lattice parameter; in order for a proper comparison to be made the 0 K lattice parameter should be
found, either from experimental data or a more extensive computational
investigation. The basis-set convergence in the correlation residual appears
to be quite slow (note the large basis-set correction in Table 4.4); this can
be corrected by calculating corrections in larger basis sets or through the
use of explicitly correlated electronic structure methods [17].

Table 4.4 Hierarchical Results for Lithium Fluoride
at the 298K Experimental Lattice Parameter, $a = 4.026$ Å

Contribution	Energy/mE_h
E^{HF}	−245.954
E^{corr}_{mol}	−77.705
$2E^{111}$ MP2/cc-pVTZ $N = 36$	−2.706
δbasis MP2/cc-pV[T,Q]Z $N = 16$	+5.254
δCCSD(T) CCSD(T)/cc-pVTZ $N = 12$	−3.846
δcore MP2/cc-pCVTZ $N = 16$	−2.904
ZPE PBE phonons	+4.851
$E^{0\,K}_{coh}$	−323.010

4.2.4   Neon

The performance of the hierarchical scheme has also been tested to determine the correlation energy of crystalline neon. Neon forms a face-centered-cubic (fcc) crystal with an empirically determined lattice parameter of 4.35 Å [18]. This value is obtained from the extrapolation to zero kelvin of the observed lattice parameter [19], and the subtraction of zero-point effects. With these approximations the empirical static cohesive energy is -1002 μE_h [20]. Using Hartree–Fock theory, the cohesive energy has previously been found to be $+685$ μE_h [21]. Therefore, the correlation contribution should amount to -1687 μE_h. It is clear that this system is dominated by correlation effects and so an accurate description of these effects is crucial.

Once again, only the correlation energy was treated with the hierarchical method [22]. With sufficiently short-ranged correlation effects the calculated value of E^{111} should converge rapidly with maximum cluster size. We have chosen DF-MP2/aug-cc-pVTZ as the reference level of theory, and the results of the hierarchical analysis are shown in Figure 4.8. Again it can be seen that there are two significant jumps in the E^{111} values, which coincide with the elimination of certain sets of clusters. At $N = 8$, chains, sheets, and parallelepipeds are all used in order to solve the equations. By $N = 12$ only sheets and parallelepipeds are used, and at $N = 28$ sheets are no longer needed.

The hierarchical scheme is extremely effective in subtracting edge effects from calculations on small clusters; in earlier work, the efficacy of the method was analyzed using a many-body expansion [22]. The many-body expansion truncated at the two-body level for the energy of a

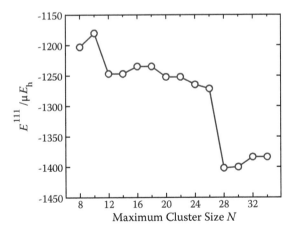

Figure 4.8 E^{111} plotted as a function of maximum cluster size N for Ne, $a = 4.35$ Å at the DF-MP2/aug-cc-pVTZ level of theory.

Table 4.5 Coefficients for Contributions at Each Possible Interatomic Distance d (in Lattice Parameter Units a) in the Crystal to E^{111} Values Computed at Particular Maximum Cluster Sizes N Using the Two-Body Approximation to Cluster Energies. It Can Be Seen That as N Increases, There Is an Increase in the Number of Coefficients That Agree with the Bulk. (For Clarity, Those that Agree Are Shown in Bold.)

d/a	$1/\sqrt{2}$	1	$\sqrt{3/2}$	$\sqrt{2}$	$\sqrt{5/2}$	$\sqrt{3}$	$\sqrt{7/2}$	2	$3/\sqrt{2}$	$\sqrt{5}$	$\sqrt{11/2}$
n_d^{12}	6	3	5	−1	2	1	−2	0	−3	0	1
n_d^{28}	**6**	**3**	12	6	12	−3/2	−15/2	**3**	−21/4	−5	15/4
n_d^{64}	**6**	**3**	**12**	**6**	**12**	**4**	**24**	**3**	**18**	**12**	−32/3
n_d^{∞}	6	3	12	6	12	4	24	3	18	12	12

cluster can be written

$$E_{\ell mn} = \sum_d n_d V(d) \tag{4.12}$$

where V is the interatomic potential, and where the summation runs over the interatomic distances d that appear $n_d > 0$ times in the cluster. By inserting two-body expansions of the energies $E_{\ell mn}$ that appear in the hierarchical equations, it is possible to compute the hierarchical cohesive energy at the two-body level, and this reveals an interesting property of the hierarchical method. In the bulk environment, a given neon atom has 12 nearest neighbors (with distance $a/\sqrt{2}$) and this contributes $6V(a/\sqrt{2})$ to the cohesive energy. Similar contributions can be calculated for all of the interatomic distances represented in the crystal

$$E_{\mathrm{coh}} = \sum_d n_d^{\infty} V(d) \tag{4.13}$$

and the first few values of n_d^{∞} are shown in the last row of Table 4.5.

When the hierarchical analysis is performed using two-body cluster energies, a cohesive energy is obtained in the same form

$$E_{\mathrm{coh}}^N = \sum_d n_d^N V(d) \tag{4.14}$$

but with different coefficients n_d^N. Interestingly, even for small values of N, the first few coefficients n_d^N agree perfectly with the values in the infinite crystal n_d^{∞}. In fact, there is agreement (i.e., $n_d^N = n_d^{\infty}$) even when none of the clusters used to solve the hierarchical equation contains an atom with n_d^{∞} neighbors at distance d. This ability of the hierarchical method to reproduce aspects of the bulk environment that are not represented in any of the input clusters offers an alternative view of the success of the method. The reproduction of the n_d^∞ coefficients at various levels of the hierarchical analysis is presented in Table 4.5.

Since the two-body contribution to the binding energy of even very large clusters can be computed cheaply, we can correct our hierarchical analysis using a two-body correction out to very large cluster sizes. For neon, the hierarchical method has been performed using clusters of up to 400 atoms in the two-body approximation. These results are presented in Figure 4.9 and a correction based on the difference in energy between DF-MP2/aug-cc-pVTZ counterpoise-corrected two-body calculations at $N = 32$ and $N = 400$ is used in the final evaluation of the cohesive energy. The data also suggests that in order to achieve a high level of accuracy for neon, quite large cluster sizes would be needed (at least 64 atoms to reduce errors to about 1%). The results of the hierarchical analysis, together with this correction for cluster size and other corrections for basis set completeness and level of theory, are shown in Table 4.6.

The best estimate of the correlation component of the cohesive energy of -1750.72 μE_h is within 100 μE_h of the experimental value, so there is clearly room for improvement. Crucially, the error could be reduced systematically by improving the basis sets used for the correction terms, by performing calculations at higher values of N, by accounting for the effects of core correlation, by including correlation beyond the CCSD(T) level of theory, and by taking account of three-body effects in the large-N correction term.

4.3   Calibration of other methods

Although the hierarchical methods appear to have wider generality, particular care was invested in the calculations on lithium hydride [7,12]. This work has provided reliable results that have already been used for benchmarking other methods. By reproducing the calculation outlined previously without high-level corrections, we have been able to produce estimates of the static cohesive energy at $a = 4.084$ Å, which can relied upon in the testing of newly developed periodic electronic structure codes. The results at the Hartree–Fock and MP2 levels are shown in Table 4.7, and other reference values are provided in our original work on this subject [12].

4.4   Conclusions

The hierarchical method makes it possible to treat electron correlation in crystalline solids using molecular electronic structure methods. The power of the approach derives from a simple but effective elimination of surface effects from calculations on finite clusters.

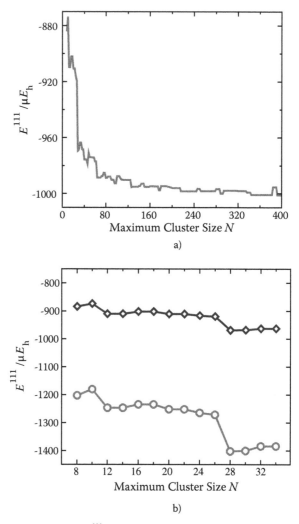

Figure 4.9 Convergence of E^{111} with increasing maximum cluster size N for *Ne*, $a = 4.35$ Å. All correlation energies were calculated using MP2/aug-cc-pVTZ. In (a), the two-body hierarchical results are shown to large maximum N; in (b) the full (circles) and two-body (diamonds) hierarchical results are compared as a function of maximum cluster sizes N. Correlation contributions to binding energies were calculated using DF-MP2/aug-cc-pVTZ.

For a simple solid, like lithium hydride, it is possible to converge the calculations to high precision with respect to both cluster size and the treatment of electron correlation, and this, coupled with highly accurate Hartree–Fock results and DFT phonon frequencies, reproduces empirical properties of the material to within experimental uncertainties.

Table 4.6 Hierarchical Result for E^{111} at $N = 34$ Using DF-MP2/aug-cc-pVTZ and Corrections to That Value Based on Calculations Using Different Methods, Basis Sets, and Cluster Sizes. All Values Quoted in μE_h.

Term	N	E^{111}
DF-MP2/AVTZ	34	−1383.68
2-body DF-MP2/AVTZ	400	−37.56
DF-MP2/AV[TQ]Z	8	−111.01
CCSD(T)/AVDZ	8	−218.47
$E_{\text{coh}}^{\text{corr}}$		−1750.72
$E_{\text{coh}}^{\text{corr}}$ experimental		−1687

AVTZ is used as an abbreviation for aug-cc-pVTZ, and AV[TQ]Z for results with cubically extrapolated correlation energies using the corresponding triple-zeta and quadruple-zeta basis sets (see T. Helgaker, W. Klopper, H. Koch, and J. Noga, *J. Chem. Phys.* 106, 9639 (1997)).

High accuracy also appears, from preliminary investigations, to be achievable for other materials.

For crystalline neon, where the cohesion is entirely dominated by dispersion, convergence of the method to high relative accuracy is demanding, as the long-range (R^{-6}) decay of the interactions makes it necessary to consider very large clusters. However, a simple two-body approximation can be used to correct for this deficiency using interatomic potentials derived from the ab initio calculations.

One can reverse the analysis: we could regard the calculation on the cohesive energy of neon (Section 4.2.4) as a many-body expansion of the cohesive energy truncated at the two-body level, with a correction for many-body effects computed using the hierarchical method. We hope that

Table 4.7 All-Electron Basis-Set Limit Estimates of the Static Cohesive Energy of Lithium Hydride. The CP2K/Gaussian result can be taken to be exact to all digits presented. All energies shown in millihartree.

Method	Hierarchical[a]	VASP[b]	CP2K[c]
HF	−131.95	−131.7	−131.949
MP2	−175.3	−175.7	

[a] Calculated at $a = 4.084$ Å at the Hartree–Fock [7] and MP2 [12] levels of theory.
[b] Calculated by Marsman et al. using VASP [24]
[c] Calculated by Paier et al. using CP2K and Gaussian [25]

this points toward a fruitful way in which the hierarchical and incremental approaches to electron correlation in solids can complement each other.

References

[1] W. M. C. Foulkes, L. Mitas, R. J. Needs, and G. Rajagopal, Quantum Monte Carlo simulations of solids, *Rev. Mod. Phys.* 73, 33 (2001).

[2] A. Lüchow, R. Petz, and T. C. Scott, Direct optimization of nodal hypersurfaces in approximate wave functions, *J. Chem. Phys.* 126, 144110 (2007).

[3] C. J. Umrigar, J. Toulouse, C. Filippi, S. Sorella, and R. G. Hennig, Alleviation of the fermion-sign problem by optimization of manby-body wave functions, *Phys. Rev. Lett.* 98, 110201 (2007).

[4] A. Badinski, P. D. Haynes, and R. J. Needs, Nodal Pulay terms for accurate diffusion quantum Monte Carlo forces, *Phys. Rev. B* 77, 085111 (2008).

[5] A. Badinski, J. R. Trail, and R. J. Needs, Energy derivatives in quantum Monte Carlo involving the zero-variance property, *J. Chem. Phys.* 129, 224101 (2008).

[6] F. R. Manby, D. Alfè, and M. J. Gillan, Extension of molecular electronic structure methods to the solid state: computation of the cohesive energy of lithium hydride, *Phys. Chem. Chem. Phys.* 8, 5178 (2006).

[7] M. J. Gillan, F. R. Manby, D. Alfè, and S. de Gironcoli, High-precision calculation of Hartree–Fock energy of crystals, *J. Comput. Chem.* 29, 2098 (2008).

[8] A. Abdurahman, A. Shukla, and M. Dolg, Ab initio treatment of electron correlations in polymers: Lithium hydride chain and beryllium hydride polymer, *J. Chem. Phys.* 112, 4801 (2000).

[9] G. D. Barrera, D. Colognesi, P. C. H. Mitchell, and A. J. Ramirez-Cuesta, LDA or GGA? A combined experimental inelastic neutron scattering and ab initio lattice dynamics study of alkali metal hydrides, *Chem. Phys.* 317, 119 (2005).

[10] A. Shukla, M. Dolg, P. Fulde, and H. Stoll, Wave-function-based correlated ab initio calculations on crystalline solids, *Phys. Rev. B* 60, 5211 (1999).

[11] S. J. Binnie, E. Sola, D. Alfè, and M. J. Gillan, Benchmarking DFT surface energies with quantum Monte Carlo, *Molecular Simulation* 35, 609 (2009).

[12] S. J. Nolan, M. J. Gillan, D. Alfè, N. L. Allan, and F. R. Manby, Calculation of properties of crystalline lithium hydride using correlated wave function theory, *Phys. Rev. B* 80, 165109 (2009).

[13] S. Casassa, M. Halo, L. Maschio, C. Roetti, and C. Pisani, Beyond a Hartree–Fock description of crystalline solids: The case of lithium hydride, *Theor. Chem. Acc.* 117, 781 (2007).

[14] H.-J. Werner, P. J. Knowles, R. Lindh, F. R. Manby, M. Schütz, et al., MOLPRO, Version 2008.1, a package of ab initio programs, 2008, see http://www.molpro.net.

[15] V. Fiorentini and M. Methfessel, Extracting convergent surface energies from slab calculations, *J. Phys.: Condens. Matter* 8, 6525 (1996).

[16] *NIST-JANAF thermochemical tables*, 4th ed., edited by M. W. Chase, Jr. *J. Phys. Chem. Ref. Data* Monograph No. 9, (1998).

[17] W. Klopper, F. R. Manby, S. Ten-no, and E. F. Valeev, R12 methods in explicitly correlated molecular electronic structure theory, *Int. Rev. Phys. Chem.* 25, 427 (2006).

[18] K. Rosciszewski, B. Paulus, P. Fulde, and H. Stoll, Ab initio calculation of ground-state properties of rare-gas crystals, *Phys. Rev. B* 60, 7905 (1999).

[19] D. N. Batchelder, D. L. Losee, and R. O. Simmons, Measurements of lattice constant, thermal expansion, and isothermal compressibility of neon single crystals, *Phys. Rev.* 162, 767 (1967).

[20] G. McConville, New values of sublimation energy L_0 for natural neon and its isotopes, *J. Chem. Phys.* 60, 4093 (1974).

[21] M. Halo, S. Cassassa, L. Maschio, and C. Pisani, Periodic local-MP2 computational study of crystalline neon, *Phys. Chem. Chem. Phys.* 11, 568 (2009).

[22] S. J. Nolan, P. J. Bygrave, N. L. Allan, and F. R. Manby, Comparison of the incremental and hierarchical methods for crystalline neon, *J. Phys.: Cond. Matt.* 22, 074201 (2010).

[23] T. Helgaker, W. Klopper, H. Koch, and J. Noga, Basis-set convergence of correlated calculations on water, *J Chem. Phys.* 106, 9639 (1997).

[24] M. Marsman, A. Grüneis, J. Paier, and G. Kresse, Second-order Møller–Plesset perturbation theory applied to extended systems. I. Within the projector-augmented-wave formalism using a plane wave basis set, *J. Chem. Phys.* 130, 184103 (2009).

[25] J. Paier, C. V. Diaconu, G. E. Scuseria, M. Guidon, J. VandeVondele, and J. Hutter, Accurate Hartree–Fock energy of extended systems using large Gaussian basis sets, *Phys. Rev. B* 80, 174114 (2009).

chapter five

Electrostatically embedded many-body expansion for large systems

Erin Dahlke Speetzen, Hannah R. Leverentz, Hai Lin,
and Donald G. Truhlar

Contents

5.1 Introduction

The need to carry out accurate atomistic simulations of condensed-phase systems is a major challenge to computational chemists. For large systems (often tens or hundreds of thousands of atoms) quantum chemistry is usually too expensive to be practical, and more approximate methods such as molecular mechanics force fields have been utilized instead. Molecular mechanics force fields involve classical mechanical force constants and classical electrostatics, but they do not invoke quantum mechanical wave functions and usually do not involve polarizability. Thus they have the advantage of being easy to implement and efficient to compute, but there are many situations in which the use of molecular mechanics force fields is not applicable or not reliable. For example, most molecular mechanics force fields do not allow for the making and breaking of bonds, and so these methods cannot be used to examine chemical reactions. Furthermore, while these methods may give good agreement with experiments for physical properties against which they have been parameterized, they often do not do well when tested on systems or properties outside of the parameterization set.

With increases in computer power, classical potentials have been replaced in many cases by the use of density functional theory (DFT) [1,2], which in some cases [3,4] is able to achieve accuracy comparable to or better than highly correlated levels of wave function theory (WFT), [5–7], such as coupled cluster theory with single and double excitations [8] (CCSD), or coupled cluster theory with single and double excitations and quasipertur-bative connected triple excitations [9] (CCSD[T]), while maintaining much more favorable scaling with respect to system size. However, despite these advances the desire to be able to apply correlated levels of wave function theory (such as second-, third-, or fourth-order Møller–Plesset perturbation theory [10–13] or coupled cluster theory) to extended systems remains a goal for many computation researchers.

One of the largest problems standing in the way of this goal is the scaling of such methods with respect to increasing system size. The most basic WFT method, Hartree–Fock theory, has a computational effort that scales as N^4 where, throughout the whole chapter, N is the number of atoms in the system [5]. (Hybrid DFT also scales as N^4.) Hartree–Fock methods have been applied to very large systems; however, even though Hartree–Fock theory fully accounts for the exchange energy of the electrons of the same spin, the lack of dynamical correlation energy means that it does not provide sufficient accuracy for most applications of chemical interest. In order to obtain the accuracy needed, one should use correlated levels of electronic structure theory, but conventional correlated WFT calculations suffer from high-order scaling of the computational cost with respect to increasing system size and are generally impractical for very large systems. For example [5], MP2 scales as N^5, MP3, CISD, MP4SQD, CCSD, and QCISD scale as N^6, and MP4, CCSD(T), and QCISD(T) scale as N^7. The high-order scaling of these methods, as originally developed, has made them too expensive to be practical for use on extended systems without introducing strategies for coping with large size.

One way to overcome the expensive scaling is to introduce localized orbitals and computationally screen out excitations that involve widely spatially separated orbitals [14] (see also Chapters 1, 2, and 3). Another way to circumvent these problems is the introduction of fragment-based methods. Fragment-based methods are built on the assumption that one can divide the system into a set of monomers (which may be molecular fragments, single molecules, or collections of molecules) and express the total energy as a combination of the energies of the monomers, dimers, trimers, etc. These methods have the benefit of reducing a single very expensive calculation to a large number of small calculations. However, the simplest of these methods, the many-body method, does not give good quantitative results until one includes third-order terms (for moderate accuracy) or higher-order terms (for high accuracy), and the cost to obtain the desired accuracy may be prohibitive. In order to increase the accuracy of these calculations at a low order (i.e., including only dimers or trimers) several groups have tried modifying these methods to include the effects of the electrostatic potential of the other fragments in the system, and a variety of fragment-based approaches have been proposed in the literature [15–27]. (See also Chapters 3, 6, and 7). The emphasis here is on our own methods for carrying out these types of calculations: first, the electrostatically embedded many-body (EE-MB) method [26], which is based on the many-body method but is simpler to implement than other fragment-based methods such as the fragment molecular orbital (FMO) method or the electrostatic-field-adapted molecular-fragment-with-conjugated-caps (EFA-MFCC) method, and second, the electrostatically embedded many-body expansion of the correlation energy (EE-MB-CE) [27].

To apply fragment methods to general systems, one must cut covalent bonds to make fragments, and then the dangling bonds must be capped by link atoms or more complicated strategies [15, 16, 18–23]. This is an important research goal, but the present article is focused mainly on systems where the division into fragments does not require cutting bonds. This includes molecular clusters, molecular liquids, and molecular solids.

5.2   *Many-body methods*

Consider a system of M interacting units, to be called monomers. These monomers may be covalently connected or noncovalently connected; however, as stated at the end of the introduction, in the present chapter we will focus mainly on monomers that are not covalently connected, and the equations apply only to this case. The total energy of this system, E, can be written, without approximation, as a series of m-body potentials

$$E = V_1 + V_2 + V_3 + \cdots + V_M \tag{5.1}$$

where

$$V_1 = \sum_{i}^{M} E_i \tag{5.2}$$

$$V_2 = \sum_{i<j}^{M} (E_{ij} - E_i - E_j) \tag{5.3}$$

$$V_3 = \sum_{i<j<k}^{M} [(E_{ijk} - E_i - E_j - E_k) - (E_{ij} - E_i - E_j) - (E_{ik} - E_i - E_k)$$
$$-(E_{jk} - E_j - E_k)] \tag{5.4}$$

and higher-order terms are defined analogously. In these equations E_i represents the energy of each of the M monomers in the system, E_{ij} the energy of each of the $M(M-1)/2!$ dimers in the system, E_{ijk} the energy of each of the $M(M-1)(M-2)/3!$ trimers in the system, and so forth.

In the limit to which this expansion is carried, the Mth order, the energy returned is the total energy of the system. In order to invoke the many-body approximation, one makes the choice to truncate this series instead at the mth order (where $m < M$). If one chooses to truncate Equation (5.1) after the second term ($m = 2$), the total energy of the system is approximated as

$$E_{PA} = \sum_{i<j}^{M} E_{ij} - (M-2)\sum_{i}^{M} E_i \qquad (5.5)$$

where E_{PA} denotes the pairwise additive energy, and E_{ij} and E_i retain the same meanings as above. If one chooses to truncate after the third term ($m = 3$) the total energy of the system is approximated as

$$E_{3B} = \sum_{i<j<k}^{M} E_{ijk} - (M-3)\sum_{i<j}^{M} E_{ij} + \frac{(M-2)(M-3)}{2}\sum_{i}^{M} E_i \qquad (5.6)$$

where E_{3B} denotes the three-body approximation to the total energy, and E_{ijk}, E_{ij}, and E_i retain the same meanings as above.

The benefit of making such an approximation is that for a large system it reduces a single large and expensive calculation to a large number of small and affordable calculations. Moreover, since each of these small calculations is independent of the others, each one may be run on a different processor, leading to a calculation that can easily be parallelized on a large number of processors. The accuracy of the energy depends on how many m-body terms are retained. While the pairwise additive method may provide qualitatively correct results, if one hopes to obtain quantitative accuracy, one must include higher-order terms, particularly if the monomers are known to have large intermolecular interactions, such those that occur among water molecules [28].

One obvious way to avoid the problem mentioned above would be to continue to include higher-order terms until a satisfactory level of accuracy is obtained. However, for large systems the number of such higher-order terms rises quickly. Table 5.1 shows the number of clusters (dimers through hexamers) that one would need to calculate for a system with 64 monomers. As seen in this table, the number of clusters one must calculate rises rapidly with increasing cluster size (roughly as $M^m/m!$). Moreover, the cost of each of these calculations rises, so that not only must you calculate a much larger number of these higher-order terms, but as the size of the cluster increases, so does the cost to calculate its energy.

Table 5.1 Number of Clusters to Calculate
for Cluster Sizes $m = 1 - 6$ for $M = 64$

Cluster size (m)	Number of clusters
Monomer (1)	64
Dimer (2)	2 016
Trimer (3)	41 664
Tetramer (4)	635 376
Pentamer (5)	7 624 512
Hexamer (6)	74 974 368

In order to consider the cost associated with carrying out a many-body expansion it is useful to compare the amount of time it would take to calculate the energy of the system with and without invoking the many-body approximation. The calculation of the energy without using the many-body expansion (i.e., a calculation of the energy of a supersystem containing all M monomers in the system at a given level of theory and with the given basis set) will be referred to as a *conventional calculation*. The calculation of energy of the same system using the pairwise additive approximation will be denoted PA, while the calculation using the three-body approximation will be denoted 3B. Table 5.2 shows a comparison of the theoretical timings for the calculation of the energies computed in the conventional manner and with the pairwise additive and three-body approximations, for methods that scale as $a N^5$, $b N^6$, and $c N^7$, a, b, and c are unknown prefactors specific to each level of electronic structure theory and basis set. One can see that even on a single processor the many-body methods are far more cost effective than conventional calculations. The pairwise additive calculations are between 5 and 7 orders of magnitude faster than a conventional calculation, while the three-body calculations are between 2 and 5 orders of magnitude faster. In fact, for the methods shown in Table 5.2, one may go up to the 5th order in the many-body expansion before the cost of the many-body expansion surpasses the cost of the conventional calculation.

Table 5.2 Comparison of Hypothetical Timings for Full
Calculations and Many-Body Calculations for a System
Containing 64 Monomers

Scaling	Conventional	PA	3B
$a N^5$	$1.1 \times 10^9 a$	$6.5 \times 10^4 a$	$1.0 \times 10^7 a$
$b N^6$	$6.9 \times 10^{10} b$	$1.3 \times 10^5 b$	$3.1 \times 10^7 b$
$c N^7$	$4.4 \times 10^{12} c$	$2.5 \times 10^5 c$	$9.1 \times 10^7 c$

5.3 Electrostatically embedded many-body methods

5.3.1 EE-MB

The electrostatically embedded many-body (EE-MB) methods have been presented previously [26], and a summary of the methods is given here. The EE-MB methods are an extension of the basic many-body idea. The key difference is that in the basic many-body methods, the energy of each monomer, dimer, or trimer is calculated in vacuum; in the EE-MB methods each monomer, dimer, or trimer is calculated in a field of point charges centered on the nuclear positions of the missing monomers. For example, in the electrostatically embedded pairwise additive (EE-PA) method, the energy of the system is written as

$$E_{EE-PA} = \sum_{i<j}^{M} E'_{ij} - (M-2) \sum_{i}^{M} E'_{i} \qquad (5.7)$$

where E'_i are the energies of each monomer embedded in a field of nuclear-centered point charges representing the other $M-1$ monomers and E'_{ij} are the energies of each dimer embedded in a field of nuclear-centered point charges representing the other $M-2$ monomers. Similarly, in the electrostatically embedded three-body method, the energy of the system is written as

$$E_{EE-3B} = \sum_{i<j<k}^{M} E'_{ijk} - (M-3) \sum_{i<j}^{M} E'_{ij} - \frac{(M-2)(M-3)}{2} \sum_{i}^{M} E'_{i} \qquad (5.8)$$

where E'_i and E'_{ij} have the same meaning as above and E'_{ijk} are the energies of each trimer embedded in a field of nuclear-centered point charges representing the other $M-3$ monomers.

Just as with the unembedded many-body expansions, if one does not truncate the EE-MB series the energy obtained is the total energy of the system. The presence of the point charges in the EE-MB method is meant to increase the rate of convergence of Equation (5.1) by incorporating the higher-order many-body effects in an average way. Work on clusters containing pure water [26–28] as well as mixed clusters containing ammonia and water [29], and water, ammonia, sulfuric acid, and ammonium and bisulfate ions [30] have shown the accuracy of the method to be relatively independent of the choice, within reason, of partial atomic charges (and if the expansion is not truncated the result is completely independent of the type of charges used), although in general it has been found that larger charges produce slightly better results. For simplicity, calculations have been carried out on gas-phase monomers (i.e., for water, a water molecule with an O-H bond length of 0.9572 Å and an H-O-H bond angle of 104.52°)

to determine the partial atomic charges for that species. One may use a variety of charge models including Mulliken, ChelpG, or CM4 charges, and those charges in turn are used to represent the missing monomers in the EE-MB calculations.

Recently, the sensitivity of the accuracy of the EE-MB methods has been examined with respect to a wide variety of charge models, including models in which the charges depend on the geometry of the system being studied [30]. While it was found that the use of these geometry-dependent charges did improve the accuracy of the methods very slightly, the small gain in accuracy is not worth the additional cost of calculating the geometry-dependent charges, and so it is recommended to use charges obtained from gas-phase monomers with the method and charge representation of your choice.

The process for calculating the EE-MB energy can then be summarized as follows:

1. Calculate the embedding charges for each monomer type in the system using gas-phase molecules and the electronic structure method and charge model of your choice.
2. For each m-body cluster in the expansion (for EE-PA $m = 1$ and 2; for EE-3B $m = 1$, 2, and 3) embed the cluster in a field of nuclear-centered point charges (as determined in step one) representing the other $M - m$ monomers and calculate the energy of each embedded cluster with the electronic structure method and basis set of your choice.
3. Calculate the total energy of the system using either Equation (5.7) (EE-PA) or Equation (5.8) (EE-3B).

5.3.2   EE-MB-CE

In practice, the EE-MB methods described in the previous section can be used in conjunction with any level of electronic structure theory including both WFT and DFT. However, an additional approximation can be made when using correlated levels of electronic structure theory such as MP2, CCSD, or CCSD(T). When a post-Hartree–Fock correlated level of theory is used the total energy can be written as

$$E_X = E_{HF} + \Delta E_{corr,X} \tag{5.9}$$

where E_X is the electronic energy of correlated method X (where X = MP2, CCSD, CCSD(T), etc.), E_{HF} is the Hartree–Fock energy of the system, and $\Delta E_{corr,X}$ is the correlation energy for method X. Because the total energy is a linear combination of the correlation energy and the Hartree–Fock energy, and since each term, V_m, in the many-body expansion is a linear combination of energies for the 1- to m-body clusters, we can use Equation (5.9) to write V_m as

$$V_m = V_{m,\text{HF}} + \Delta V_{m,\text{corr}}. \tag{5.10}$$

A consequence of Equation (5.10) is that the energy of the system can be written as the sum of two different many-body expansions: one for the expansion of the Hartree–Fock energy and one for the expansion of the correlation energy, that is,

$$V = (V_{1,\text{HF}} + V_{2,\text{HF}} + V_{3,\text{HF}} + \cdots) + (\Delta V_{1,\text{corr}} + \Delta V_{2,\text{corr}} + \Delta V_{3,\text{corr}} \cdots). \tag{5.11}$$

It is not unreasonable to treat the expansion of the Hartree–Fock energy and the expansion of the correlation energy differently. For chemical systems without a significant amount of static correlation the Hartree–Fock energy often dominates even the changes in the electronic energy of the system, and therefore one expects that changes in Hartree–Fock energy will often be larger than changes in correlation energy for the process of interest, for example, for a binding event. In such cases, much of the error associated with the truncation of the EE-MB expansion is due to the truncation of the Hartree–Fock energy and not the correlation energy. This is not to imply that the error from the truncation of the correlation energy is not important, only that this error is likely to be considerably smaller, at a given order of expansion, than the error from truncating the Hartree–Fock energy. In such cases, one might truncate the expansion of the correlation energy at a lower level than the expansion of the Hartree–Fock energy.

The motivation for EE-MB-CE is strictly practical, based on the effect that such a procedure has on the cost of the calculation. Since Hartree–Fock theory has more favorable scaling than post-Hartree–Fock methods, when considering higher-order terms (such as three-body or four-body terms) one can afford to calculate larger clusters with Hartree–Fock theory than would be practical with post-Hartree–Fock methods. For moderate-sized systems (up to a few hundred atoms) one can, if desired, calculate the Hartree–Fock energy of the entire system (i.e., carry the many-body expansion to M^{th} order) and carry out a many-body expansion of only the correlation energy [27].

5.4   Performance

The accuracies of the EE-MB and EE-MB-CE methods were examined in [27] for a series of pure water clusters, ranging in size from 5 to 20 water molecules and taken from the Cambridge Cluster Database [31], at the MP2/jul-cc-pVTZ level of theory (where jul- denotes semidiffuse, so

Table 5.3 Comparison of Mean Errors
(kcal/mol) at the MP2/jul-cc-pVTZ Level
of Theory for Pure Water Clusters $(H_2O)_n$
with $n = 5 - 20$

	MSE	MUE	RMSE
PA	15.95	15.95	17.55
3B	0.55	0.56	0.71
EE-PA	0.80	0.80	0.84
EE-3B	−0.34	0.35	0.51
PA-CE	0.22	0.22	0.24
3B-CE	−0.05	0.17	0.24
EE-PA-CE	−0.10	0.10	0.11
EE-3B-CE	−0.23	0.23	0.34

the jul-cc-pVTZ basis set uses the aug-cc-pVTZ basis set [32,33] on oxygen and the cc-pVTZ basis set [34] on hydrogen). Eight different many-body methods were used: PA, 3B, EE-PA, EE-3B, PA-CE, 3B-CE, EE-PA-CE, and EE-3B-CE; each method has been described in Section 5.2 or 5.3. The energy calculated with each of these methods was compared to the energy of a conventional calculation on the same cluster at the same level of theory. The mean errors for each of the many-body methods, calculated over the complete data set, are shown in Table 5.3. The key results can be summarized as follows:

1. The inclusion of point charges dramatically improves the accuracy of the pairwise additive approximation. A comparison of the results of the PA method to the EE-PA method shows that the mean unsigned error is reduced from 15.95 kcal/mol to 0.80 kcal/mol. The average binding energy for these clusters was 105.48 kcal/mol, so an error of 0.80 kcal/mol corresponds to 0.8% of the average binding energy.
2. Even without inclusion of the embedding charges, the inclusion of the full Hartree–Fock energy with a pairwise additive treatment of the correlation energy (i.e., comparing PA to PA-CE) reduces the error by a factor of 78. The inclusion of embedding charges (i.e., comparing EE-PA to EE-PA-CE) shows an improvement of a factor of 8 in these errors. This gives errors that are as low as 0.1% of the average binding energy of these clusters by only including two body terms.

In an effort to make these calculations even more affordable (a more detailed discussion of the costs of these methods is in Section 5.6) [27] we also considered the effect of several different cutoffs on the calculation of the correlation energy. Correlation is typically short-ranged in comparison to Hartree–Fock theory, so for monomers separated by large distances it is

Table 5.4 Comparison of Mean Errors (kcal/mol) for PA-CE and EE-PA-CE
Methods with a Cutoff of 6 Å and with No Cutoff

	6 Å			No Cutoff		
	MSE	MUE	RMSE	MSE	MUE	RMSE
PA-CE	0.38	0.38	0.46	0.22	0.22	0.23
EE-PA-CE	0.02	0.07	0.09	0.09	0.09	0.10

likely that the correlation would go to zero. A comparison of the mean errors for the PA-CE and EE-PA-CE methods with no cutoff and a cutoff of 6 Å is shown in Table 5.4. These results show that with a cutoff of 6 Å, one retains the accuracy of no cutoff for the EE-PA-CE method. The implications of this result on the cost of these calculations will be considered in the next section.

To assess the accuracy of the EE-MB and EE-MB-CE methods for other levels of correlated electronic structure theory, the relative energies of a series of low-lying water hexamers were used. The water hexamers were chosen for two reasons:

1. Due to the cost of the post-MP2 methods, which scale as N^6 and N^7, we were limited to clusters on the order of 5 heavy atoms.
2. There are 5 different water hexamers that lie within 3 kcal/mol of each other, making this system a good test of the capabilities and accuracies of these methods.

The following many-body methods were used: PA, PA-CE, EE-PA, EE-PA-CE, 3B, 3B-CE, EE-3B, and EE-3B-CE with the following levels of wavefunction theory: HF, MP2, MP3, MP4D, MP4DQ, MP4SDQ, MP4, CCSD, and CCSD(T) with the jul-cc-pVTZ basis set, that is, the aug-cc-pVTZ basis set on oxygen and the cc-pVTZ basis set on hydrogen. In the following discussion of errors, we are always referring to the error relative to a full calculation at the same level of theory. We found that the error of the many-body method was largely independent of the level of wave function theory used. Table 5.5 presents the average mean unsigned error (relative to the conventional calculation at each level of theory) and standard deviation over all nine levels of electronic structure theory for each of the many body methods. Examining this table shows that all of these methods have a very low standard deviation, indicating that the errors associated with the

Table 5.5 Average Mean Unsigned Errors (kcal/mol) and Standard Deviations
for Many-Body Methods and Conventional Calculations at the Same Level of
Theory for Water Hexamers

	PA	PA-CE	EE-PA	EE-PA-CE	3B	3B-CE	EE-3B	EE-3B-CE
Ave. MUE	11.77	0.10	1.00	0.14	1.24	0.16	0.12	0.03
Std. Dev.	0.06	0.01	0.05	0.05	0.03	0.03	0.01	0.01

many-body methods are consistent over all levels of electronic structure theory tested. This indicates that conclusions drawn about the accuracies of the many-body method at more affordable levels of theory (e.g., MP2) may be extrapolated to more expensive levels of theory (e.g., CCSD(T)).

If one is interested in trying to extrapolate results from one level of electronic structure theory to another, it bears mentioning that we have found that the results obtained by using the EE-MB methods with density functional theory have accuracies similar to those obtained when MP2 is used. While the focus of this review is on methods that are able to accurately calculate correlation energies for condensed-phase systems, the lower scaling of DFT (which scales as N^3 for local functionals, i.e., functionals without Hartree–Fock exchange, and N^4 for hybrid functionals, i.e., functionals with a nonzero percentage of Hartree–Fock exchange, compared to MP2, which scales as N^5) allows one to test the EE-MB expansions against larger and more complex clusters than would be possible using correlated levels of wave function theory. For example, using DFT, the EE-PA and EE-3B methods have been applied to systems containing 64 water molecules [35]; this would be out of reach even for MP2, which has the lowest scaling of any of the post-Hartree–Fock methods. Additionally, we have applied DFT to the study of mixed clusters containing ammonia and water [29] and also clusters containing water, ammonia, sulfuric acid, and ammonium and bisulfate ions [30]. In the case of the ammonia–water clusters we studied the relative energies of a series of five tetramers ($NH_3(H_2O)_3$), five pentamers ($NH_3(H_2O)_4$), and two hexamers ($NH_3(H_2O)_5$) and found that on average, the MP2 results had a slightly lower mean unsigned error over the data set than the DFT methods tested. We also found that the average difference between the mean unsigned error for the DFT results and the MP2 results never exceeded 0.1 kcal/mol per monomer at either the EE-PA level or the EE-3B level of theory. We have also applied the EE-PA and EE-3B methods to the study of water–ammonia–sulfuric acid clusters. In this study we compared the accuracy of the EE-PA and EE-3B methods over eight different clusters (containing either one or two sulfuric acid molecules, one ammonia molecule, and one to six water molecules) with three different basis sets and five different charge models. We found that the relative errors of the EE-PA methods were, in most cases, approximately 5% of the average binding energy of the clusters, and the relative errors of the EE-3B methods were less than 1% of the average binding energy of the clusters. When the relative absolute error was averaged over all eight configurations, three basis sets, and five charge models we found that the EE-PA methods had an average relative absolute error of 3.0% and the EE-3B method had an average relative absolute error of 0.3%. This study is of particular importance because it represents the first test of the EE-MB methods in systems in which ions and charge-transfer complexes were present, which may be important for condensed-phase studies.

5.5   Cost

One of the benefits of the EE-MB methods discussed in Section 5.3 is that they provide an increased accuracy over the traditional many-body methods discussed in Section 5.2, without increasing the costs shown in Table 5.5. While it is possible that the addition of the point charges to the system may affect the convergence of the self-consistent-field(SCF) iterations and therefore affect the cost of the calculation, in practice this change in cost is negligible, and so for a good approximation the cost of the EE-MB calculations is the same as the cost of a traditional many-body calculation. In the limit of large system size, the EE-PA method scales as M^2 and the EE-3B method scales as M^3 where M is the number of monomers in the system. This feature shows the promise of the EE-MB methods for use in conjunction with MP2 or CCSD(T), which have much less favorable scaling.

In considering the cost of the EE-MB-CE method, one must consider that these calculations may be done in multiple ways, depending on the order to which the expansion of the Hartree–Fock energy is carried out. For any size cluster in which the correlation energy is being calculated by post-Hartree–Fock methods (which, except for DFT, are the most common methods at present for calculating correlation energy), the Hartree–Fock energy must be calculated first and so in that respect the Hartree–Fock energy is obtained at no extra cost. The only additional cost above that of the EE-MB methods is the cost to calculate the Hartree–Fock energies for higher-order terms than are considered for the correlation energy. The extreme case would be a calculation in which the expansion of the Hartree–Fock energy is carried out to Mth order (i.e., a conventional Hartree–Fock calculation is carried out on the system). If one considers the cost to carry out an EE-PA-CE calculation at the MP2 level of theory for a system with 64 monomers, one will see that the cost to calculate the correlation energy is two orders of magnitude larger than the cost to calculate the Hartree–Fock energy of the system. Therefore, it is safe to assume that the cost of these calculations can be estimated to a good extent by the cost of the EE-PA and EE-3B calculations at the same level of theory.

The cost of the EE-MB and EE-MB-CE can be further reduced by the use of a cutoff distance in the dimer and trimer calculations. As discussed in Section 5.4, a cutoff can be used to reduce the number of dimer calculations that must be considered. In the case presented previously, approximately 44% of all dimers were able to be excluded, which would reduce the cost of the calculation by nearly the same percentage. While the precise value of the cutoff used would likely differ for each system, the time required to find this cutoff would be short in comparison to the time saved over the course of the simulation. While no study of the effect of a cutoff in the trimer calculations has been completed yet, it is likely that one could implement a similar cutoff to further reduce the cost of the calculation.

The last consideration in the cost of the EE-MB methods is the use of parallelization. Since the calculation of each monomer, dimer, and in the case of the EE-3B method, trimer, is completely independent of all other calculations, each could, in theory, be run on a different processor. The ability to spread these calculations out evenly over a large number of processors makes these calculations attractive.

5.6 Use in simulations

5.6.1 Routes for extending EE-MB to the bulk

In order to treat liquids or solids, it is necessary to eliminate surface effects by a choice of boundary conditions, such as stochastic boundary conditions, extended-wall boundary conditions, or periodic boundary conditions [36–38]. With periodic boundary conditions, each monomer interacts with all the monomers and embedding charges in all the replicated unit cells, including its own image, until convergence or until a cutoff distance is reached. For long-range forces, the sum over these interactions is only conditionally convergent, and so care must be used to obtain a physical result; a cutoff can cause artifacts. The long-range forces are more important for some properties than for others [39]; for example, short-range structure and dynamics may depend primarily on short-range forces. Various methods such as reaction fields, multipole summation, or Ewald summation can be used to sum the infinite series of interactions or approximate their effect [36–40]. For liquids the periodicity of the images is artificial, and so some workers consider it more appropriate to employ the nearest image convention in which a given monomer interacts only with the nearest from among another given monomer or its various images [36]; however, it has been recommended that while this might be acceptable in Monte Carlo simulations, it should never be used in molecular dynamics simulations [38]. A common choice for a cutoff distance, for a cubic unit cell, is less than or equal to half the width of the cell. When this is applied to a many-body treatment, one should therefore also screen the oligomer contributions and include only those where no two monomers are separated by more than the cutoff distance. This eliminates ambiguities in the three-body contributions as well as the two-body ones [41].

The application of periodic boundary conditions to EE-MB energies can be carried out in much the same way as it is carried out in simulations involving quantum mechanical/molecular mechanical (QM/MM) calculations. In this respect, the largest difference between EE-MB and QM/MM calculations is that care must be taken in an EE-MB calculation in choosing the correct dimer and trimer combinations to include in the energy calculation. When long-range interactions (such as Coulomb or dipole–dipole interactions) are not cut off, one might require an adaptive scheme [42] to switch between the original dimer or trimer and one involving images when the distance of given monomer to an interacting monomer becomes larger than the distance between one of the monomers and the image of the other.

Ewald sums are commonly used with molecular mechanics potentials, but applying them to EE-MB is more similar to using them with combined QM/MM potentials. The total energy for a QM/MM calculation can be written as

$$E(QM/MM) = E(MM) + E(QM) + V(QM/MM) \qquad (5.12)$$

where $E(MM)$ is the energy of the molecular mechanics region, $E(QM)$ is the energy of the quantum mechanical region, and $V(QM/MM)$ is the energy due to the interaction between the QM and MM regions. Calculations on the QM region are usually carried out with a background charge distribution of molecular mechanics charges. This is called electronic embedding and is similar to the type of embedding done in the EE-MB methods; and therefore each embedded monomer, dimer, or trimer calculation can be seen as a simplified QM/MM calculation where the embedding charges are the molecular mechanics region. A difference between a QM/MM calculation and an EE-MB calculation is that an EE-MB calculation does not include the interactions of the point charges with one another. Thus one needs $V(QM/MM)$ but not $E(MM)$. There are a number of examples in the literature in which periodic boundary conditions have been applied to QM/MM calculations, but almost all of these [43–46] are for semiempirical QM methods. Recently an algorithm suitable for ab initio QM methods has been presented [47].

Another way to sum the infinite series of interactions in periodic ab initio QM calculations is the fast multipole method (FMM), which can handle non-cubic unit cells [48, 49].

Another way to account for the effect of a bulk solvent on a simulated active site is provided by a class of solvent boundary potentials, which have also been applied to QM/MM calculations and which could be reformulated for EE-MB calculations [50–52].

Motivated in part by the hybrid QM:QM scheme of Sauer and coworkers [53,54], we propose another approach to extending the many-body expansion to solids. Their scheme involves a periodic DFT calculation with a local functional on the extended system and a correction to higher-level WFT based on a series of larger and larger mechanically embedded clusters; this avoids evaluating Hartree–Fock exchange on the periodic extended system, which is very expensive. To gain this advantage in the context of the present methods, we propose a method with some similarity to the EE-MB-CE method. We call this new suggestion DFT:EE-MB-HL where HL denotes higher level. In DFT:EE-MB-HL, one would carry out a periodic

DFT calculation with a local functional for the extended system and augment this with an EE-PA or EE-3B calculation of the difference (for dimers or for dimers and trimers) between a higher-level calculation (e.g., a hybrid DFT calculation or a coupled cluster calculation) and a DFT calculation with the local functional. This approach allows one to take advantage of the many efficient periodic codes for local functionals and the expected fast convergence of an EE-MB expansion of the higher-level correction. Note that for general solids this may require the extension mentioned at the end of Section 5.1.

5.6.2   Monte Carlo simulations

The use of many-body-based methods in Monte Carlo simulations has been discussed previously by Christie and Jordan [55] and in this section we will summarize their findings for unembedded many-body methods and extend the discussion to EE-MB methods. (See also the description of such methods in Chapter 7.) A key issue in Monte Carlo calculations on clusters, liquids, and amorphous solids, is that many Monte Carlo moves involve a change in only one monomer. The present discussion is limited to that kind of move.

For a many-body method that does not involve electrostatic embedding (e.g., PA, 3B, PA-CE, 3B-CE), when one makes a Monte Carlo move in which only one molecule is displaced, the only interaction terms that must be recalculated are those involving the displaced molecule. Therefore, instead of recalculating the energies of all $M(M-1)/2$ dimers, one only needs to recalculate the energy of $M-1$ dimers. Similarly, instead of recalculating all $M(M-1)(M-2)/6$ trimers, one needs only to recalculate the energy of $(M-1)(M-2)/2$ trimers.

Without using many-body expansions, the cost of each Monte Carlo move is largely determined by the scaling with respect to increased system size of the electronic structure method used. As mentioned previously, for large systems MP2 scales as N^5, while more expensive methods such as MP4 and CCSD(T) scale as N^7. For small systems, the scaling is less severe, but in this section all discussion is carried out for systems large enough that asymptotic scaling applies. While the asymptotic limit is never actually reached in real work, the use of this limit facilitates a general discussion that illustrates the key ideas.

Table 5.6 shows a series of hypothetical timings, for the calculation of a many-body (MB) Monte Carlo move for a system with 64 monomers, for methods with costs (computational efforts) that scale as aN^5, bN^6, and cN^7, where a, b, and c are unknown prefactors for each level of electronic structure theory (the prefactor also depends on the basis set). The cost to do a pairwise additive (PA) Monte Carlo move is 6 to 9 orders of magnitude smaller than calculating the energy in the conventional manner, while the

Table 5.6 Comparison of Hypothetical Timings
for a Monte Carlo Move on a System of
64 Monomers Using the PA or 3B Methods

Scaling	Conventional	PA move	3B move
$a\,N^5$	$1.1 \times 10^9 a$	$2.0 \times 10^3 a$	$4.8 \times 10^5 a$
$b\,N^6$	$6.9 \times 10^{10} b$	$4.0 \times 10^3 b$	$1.4 \times 10^6 b$
$c\,N^7$	$4.4 \times 10^{12} c$	$8.1 \times 10^3 c$	$4.4 \times 10^6 c$

cost to do a three-body (3B) calculation is 4 to 6 orders of magnitude smaller than calculating the energy in the conventional manner. In general, for a scaling proportional to N^n, the ratio of the cost of a 3B move to a full calculation is $0.5(3^n)N^{2-n}$. Furthermore, since each of the calculations in the MB move is independent of the rest, the MB calculations may be highly parallelized and so if one has access to a large number of processors the wall clock time for the calculation may be decreased significantly. In addition, as mentioned in Sections 5.4 and 5.5, for large systems, a cutoff can be used to decrease the number of dimers and trimers calculated, eventually making the method scale linearly with the number of processors.

For a many-body method that *does* involve electrostatic embedding (e.g., EE-PA, EE-3B, etc.), individual *m*-mer energies depend not only on the coordinates of the nuclei belonging to the *m*-mer itself but also on the coordinates of the embedding charges. Thus, when any atom in the system is moved, the energy of every possible embedded *m*-mer is at least slightly affected and must in principle be recalculated in order to obtain the true EE-MB energy of the system.

However, one of the advantages of the EE-MB method is that it affords ample opportunities to explore and implement additional cost-saving approximations, especially when the EE-MB energies are used during Monte Carlo simulations, which do not require analytic gradients and therefore do not require a perfectly continuous potential energy surface as long as the "jumps" in the surface are small. We already discussed the use of spatial cutoffs to reduce the number of dimer calculations by screening out dimers or trimers where one monomer is separated by a distance greater than the cutoff. The dimers and trimers that must be recalculated because the embedding charges have moved offer additional possibilities for screening or approximation. For example, prior to performing an ab initio calculation of the energy of an embedded *m*-mer at some given step in a Monte Carlo simulation, one could estimate by first-order perturbation theory how much the movement of the embedding charge changes the *m*-mers energy. If the change is smaller than some threshold, one can accept the perturbation theory estimate rather than recalculating the energy of the *m*-mer in the new configuration of embedding charges. In this way, one could use an energy cutoff to screen out relatively expensive monomer, dimer, and trimer calculations that can be adequately treated by perturbation theory. Multilevel strategies could also be implemented to save time during Monte Carlo simulations. For example, one could perform all monomer and dimer calculations at a high level of theory and then perform all trimer calculations at some lower level of theory to obtain an estimate of the EE-3B energy. Alternatively, one could perform all *m*-mer calculations at the same level of theory but only calculate the trimer energies at certain reference steps during the Monte Carlo simulations; for intermediate steps, the EE-3B energy could be approximated by adding the difference between the EE-PA energy at the current step and that at a recent reference step to the EE-3B energy at the recent reference step. Or the change in trimers due to movement of faraway embedding charges could be calculated with a smaller basis set than is used for the strongly coupled trimers. (The EE-MB-CE method could also be considered to be a multilevel strategy.) A myriad of possibilities suggest themselves. Thus, the use of screening techniques, perturbation theory, and/or multilevel strategies can yield tremendous savings at many steps during the course of a Monte Carlo simulation without losing a significant amount of the accuracy gained by the inclusion of embedding charges.

5.6.3 Molecular dynamics

In a molecular dynamics simulation, the molecules are allowed to evolve following Newton's laws of motion. This requires calculating the forces acting on the molecules. To do this efficiently, one seeks an algorithm for analytical gradients of the potential energy surface. Since the gradient is a linear operator, applying it to the EE-PA and EE-3B energies (Equations [5.8] and [5.7]) gives [35]

$$\nabla E_{EE-PA} = \sum_{i<j}^{M} \nabla E'_{ij} - (M-2) \sum_{i}^{M} \nabla E'_{i} \qquad (5.13)$$

and

$$\nabla E_{EE-3B} = \sum_{i<j<k}^{M} \nabla E'_{ijk} - (M-3) \sum_{i<j}^{M} \nabla E'_{ij} + \frac{(M-2)(M-3)}{2} \sum_{i}^{M} \nabla E'_{i}. \qquad (5.14)$$

These equations were first presented in [35], but we later discovered a bug in the program. This bug has been corrected in MBPAC-2009 [56]. As a result, analytic gradients are available for any method that has analytic gradients for the individual *m*-mers provided that the computer program allows for the use of fractionally charged point charges to be used as psuedonuclei (sometimes called *ghost atoms* or *sparkles*) in the gradient calculation. It is important to note that all of the terms on the right-hand side of Equations (5.12) and (5.13) contribute to all of the components of the gradients. For example, $\nabla E_i'$, $\nabla E_{ij}'$, and $\nabla E_{ijk}'$ all contribute to all the gradient components of all the other monomers in the system and not just to monomers i, j, and k. A key simplification in the EE-MB methods is that the magnitudes of the point charges are fixed and therefore do not need to be updated during the course of the simulation.

As a numerical example, we consider hydrogen fluoride tetramer. The geometry used for these calculations is given in Table 5.7 and the calculations were carried out using the MBPAC-2009 program [56]. Table 5.7 shows the energy of the tetramer relative to four monomers from full calculations and from calculations by EE-PA and EE-3B at three levels of WFT. The errors in binding energies are 0.60–1.6% at the EE-PA level and reduce to 0.15–0.27% at the EE-3B level. Table 5.8 shows the unsigned errors in the gradient magnitudes as well as mean and maximum unsigned errors in the gradient components for EE-PA and EE-3B at this geometry. The gradient magnitude is 0.44–0.52 hartrees per bohr, and the mean absolute value of the 24 Cartesian components of the gradient is 0.071–0.082 hartrees per bohr, depending on the level of the quantum mechanical theory. It can be seen that, in comparison with the gradient magnitudes or gradient components, the errors are generally 3 orders of magnitude smaller in EE-PA and 4 orders of magnitude smaller in EE-3B calculations.

Table 5.7 Energies (kcal/mol) of Hydrogen Fluoride Tetramer Relative to Four Monomers from Conventional Calculations and from Many-Body Methods at the Same Level of Theory[a]

	Conventional	EE-PA	EE-3B
HF/MIDI!	29.36	29.19	29.32
MP2/cc-pVTZ	17.22	17.49	17.18
CCSD/cc-pVTZ	16.87	17.09	16.84

[a] The Cartesian coordinates (in Å) of the tetramer are

F	1.346092	1.346092	0.000000
F	−1.346092	1.346092	0.000000
F	−1.346092	−1.346092	0.000000
F	1.346092	−1.346092	0.000000
H	0.631976	1.763006	0.000000
H	−1.763006	0.631976	0.000000
H	−0.631976	−1.763006	0.000000
H	1.763006	−0.631976	0.000000

The Cartesian coordinates (in Å) of the monomer are

F	1.346092	1.346092	0.000000
H	0.631976	1.763006	0.000000

Table 5.8 Unsigned Errors in Gradient Magnitudes (hartree/bohr), Mean Unsigned Errors, and Maximum Unsigned Errors in Gradient Components (hartree/bohr) by Many-Body Methods with Respect to Conventional Calculations for Hydrogen Fluoride Tetramer. (The Geometries Are Specified in Table 5.7.)

	Gradient magnitudes	
	EE-PA	EE-3B
HF/MIDI!	2.7×10^{-3}	2.0×10^{-4}
MP2/cc-pVTZ	5.8×10^{-4}	4.0×10^{-5}
CCSD/cc-pVTZ	7.1×10^{-4}	3.0×10^{-5}

	Gradient components			
	Mean		Maximum	
	EE-PA	EE-3B	EE-PA	EE-3B
HF/MIDI!	2.0×10^{-4}	2.5×10^{-5}	5.4×10^{-4}	5.5×10^{-5}
MP2/cc-pVTZ	5.3×10^{-5}	8.6×10^{-6}	1.5×10^{-4}	2.3×10^{-5}
CCSD/cc-pVTZ	5.3×10^{-5}	6.8×10^{-6}	1.4×10^{-4}	1.9×10^{-5}

One can also write a set of equations similar to Equations (5.13) and (5.14) in which the Hessian has been applied to the EE-MB energies. As a result, the EE-MB methods will also have analytic Hessians for any correlated level of theory that has analytic Hessians. This is important for studying vibrations and phonons.

In contrast to Monte Carlo simulations, the energy of the entire system must be recalculated after each time step of a molecular dynamics calculation, so even with the use of the MB methods the cost of these calculations will be quite high for a large system. However, the scaling with system size is reduced to N^2 for EE-PA and N^3 for EE-3B even without screening, and the method offers numerous possibilities for screening out dimers and trimers to reduce the scaling eventually to linear. Since conventional hybrid DFT scales as N^4, an EE-3B CCSD(T) calculation becomes less expensive than a conventional hybrid DFT calculation for large N.

5.7 Conclusions

The ability to accurately calculate correlation energies for condensed-phase systems remains an important goal of computational scientists. The electrostatically embedded many-body (EE-MB) method and the electrostatically embedded many-body expansion of the correlation energy (EE-MB-CE) method provide promising new routes to achieve this goal. The high accuracy of these methods coupled with their simple implementation, low cost, and easy parallelization make them attractive options for use on condensed-phase systems. While the EE-MB and EE-MB-CE methods have not been tested on true condensed-phase systems yet, work on large molecular clusters shows that they are able to reproduce the binding energies of these clusters to within 1% of the true energy obtained by a conventional calculation on the system, and this review discusses how the methods can be used in conjunction with available computational methodology to make them applicable to both molecular dynamics and Monte Carlo simulations of condensed-phase systems. The implementation of the EE-MB and EE-MB-CE methods in existing simulation packages and the application of these methods to interesting condensed-phase problems provide interesting and exciting possibilities for further research.

Acknowledgments

This material is based upon work supported by the National Science Foundation under grant no. CHE07-04974. H. Lin was a long-term visitor of the 2008–2009 Mathematics and Chemistry program in the Institute of Mathematics and Its Applications at the University of Minnesota.

References

[1] P. Hohenberg and W. Kohn, Inhomogeneous electron gas. *Phys. Rev.* 136, B864 (1964).
[2] W. Kohn and L. J. Sham, Self-consistent equations including exchange and correlation effects. *Phys. Rev.* 140, A1133 (1965).
[3] Y. Zhao and D. G. Truhlar, Density functionals with broad applicability in chemistry. *Acc. Chem. Res.* 41, 157 (2008).
[4] J. Zheng, Y. Zhao, and D. G. Truhlar, The DBH24/08 database and its use to assess electronic structure model chemistries for chemical reaction barrier heights. *J. Chem. Theory Comput.* 5, 808 (2009).
[5] K. Raghavachari and J. B. Anderson, Electron correlation effects in molecules. *J. Phys. Chem.* 100, 12960 (1996).
[6] M. Head-Gordon, Quantum chemistry and molecular processes. *J. Phys. Chem.* 100, 13213 (1996).
[7] J. A. Pople, Nobel lecture: Quantum chemical models. *Rev. Mod. Phys.* 71, 1267 (1999).
[8] J. Čížek, On the use of the cluster expansion and the technique of diagrams in calculations of correlation effects in atoms and molecules. *Adv. Chem. Phys.* 14, 35 (1969).
[9] K. Raghavachari, G. W. Trucks, J. A. Pople, and M. Head-Gordon, A fifth-order perturbation comparison of electron correlation theories. *Chem. Phys. Lett.* 157, 479 (1989).
[10] R. Krishnan and J. A. Pople, Approximate fourth-order perturbation theory of the electron correlation energy. *Int. J. Quant. Chem.* 14, 91 (1978).

[11] R. Krishnan, M. J. Frisch, and J. A. Pople, Contribution of triple substitutions to the electron correlation energy in fourth order perturbation theory. *J. Chem. Phys.* 72, 4244 (1980).

[12] M. Frisch, R. Krishnan, and J. Pople, A systematic study of the effect of triple substitutions on the electron correlation energy of small molecules. *Chem. Phys. Lett.* 75, 66 (1980).

[13] G. F. Adams, G. D. Bent, and R. J. Bartlett, Calculation of potential energy surfaces for HCO and HNO using many-body methods, in *Potential energy surfaces and dynamics calculations*, edited by D. G. Truhlar, pages 133–167 (Plenum, New York, 1981).

[14] S. Saebø and P. Pulay, A low-scaling method for second order Møller–Plesset calculations. *J. Chem. Phys.* 115, 3975 (2001).

[15] K. Kitaura, E. Ikeo, T. Asada, T. Nakano, and M. Uebayashi, Fragment molecular orbital method: an approximate computational method for large molecules. *Chem. Phys. Lett.* 313, 701 (1999).

[16] V. Deev and M. A. Collins, Approximate *ab initio* energies by systematic molecular fragmentation. *J. Chem. Phys.* 122, 154102/1 (2005).

[17] S. Hirata, M. Valiev, M. Dupuis, S. S. Xantheas, S. Sugiki, and H. Sekino, *Mol. Phys.* 103, 2255 (2005).

[18] X. H. Chen and J. Z. H. Zhang, Molecular fractionation with conjugated caps density matrix with pairwise interaction correction for protein energy calculation. *J. Chem. Phys.* 125, 44903/1 (2006).

[19] N. Jiang, J. Ma, and Y. Jiang, Electrostatic field-adapted molecular fractionation with conjugated caps for energy calculations of charged biomolecules. *J. Chem. Phys.* 124, 114112/1 (2006).

[20] D. M. Fedorov and K. Kitaura, Theoretical development of the fragment molecular orbital (FMO) method. in *Modern methods for theoretical physical chemistry of biopolymers*, edited by E. B. Starikov, J. P. Lewis, and S. Tanaka, pages 3–38 (Elsevier, Amsterdam, 2006).

[21] W. Li, S. Li, and Y. Jiang, Generalized energy-based fragmentation approach for computing the ground-state energies and properties of large molecules. *J. Phys. Chem. A* 111, 2193 (2007).

[22] D. G. Fedorov and K. Kitaura, Extending the power of quantum chemistry to large systems with the fragment molecular orbital method. *J. Phys. Chem. A* 111, 6904 (2007).

[23] W. Xie, L. Song, D. G. Truhlar, and J. Gao, The variational explicit polarization potential and analytical first derivative of energy: Towards a next generation force field. *J. Chem. Phys.* 128, 234108/1 (2008).

[24] S. Hirata, Fast electron-correlation methods for molecular crystals: An application to the α, β_1, and β_2 modifications of solid formic acid. *J. Chem. Phys.* 129, 204104/1 (2008).

[25] E. E. Dahlke and D. G. Truhlar, Assessment of the pairwise additive approximation and evaluation of many-body terms for water clusters. *J. Phys. Chem. B* 110, 10595 (2006).

[26] E. E. Dahlke and D. G. Truhlar, Electrostatically embedded many-body expansion for large systems, with applications to water clusters. *J. Chem. Theory Comput.* 3, 46 (2007).

[27] E. E. Dahlke and D. G. Truhlar, Electrostatically embedded many-body correlation energy, with applications to the calculation of accurate second-order Møller–Plesset perturbation theory energies for large water clusters. *J. Chem. Theory Comput.* 3, 1342 (2007).

[28] M. J. Elrod and R. J. Saykally, Many-body effects in intermolecular forces. *Chem. Rev.* 94, 1975 (1994).

[29] A. Sorkin, E. E. Dahlke, and D. G. Truhlar, Application of the electrostatically embedded many-body expansion to microsolvation of ammonia in water clusters. *J. Chem. Theory Comp.* 4, 683 (2008).

[30] H. R. Leverentz and D. G. Truhlar, Electrostatically embedded many-body approximation for systems of water, ammonia, and sulfuric acid and the dependence of its performance on embedding charges. *J. Chem. Theory Comp.* 5, 1573 (2009).

[31] D. J. Wales, J. P. K. Doye, A. Dullweber, M. P. Hodges, F. Y. Naumkin, F. Calvo, J. Hernández-Rojas, and T. F. Middleton, The Cambridge cluster database, http://www-wales.ch.cam.ac.uk/CCD.html, 2006, accessed March 8, 2006.

[32] R. A. Kendall, T. H. Dunning Jr., and R. J. Harrison, Electron affinities of the first-row atoms revisited. Systematic basis sets and wave functions. *J. Chem. Phys.* 96, 6796 (1992).

[33] D. E. Woon and T. H. Dunning Jr., Gaussian basis sets for use in correlated molecular calculations. III. The atoms aluminum through argon. *J. Chem. Phys.* 98, 1358 (1993).

[34] T. H. Dunning Jr., Gaussian basis sets for use in correlated molecular calculations. I. The atoms boron through neon and hydrogen. *J. Chem. Phys.* 90, 1007 (1989).

[35] E. E. Dahlke and D. G. Truhlar, Electrostatically embedded many-body expansion for simulations. *J. Chem. Theory Comput.* 4, 1 (2008).

[36] W. F. van Gunsteren, Molecular dynamics and stochastic dynamics: A primer, in *Computer simulation of biomolecular systems*, edited by W. F. van Gunsteren, P. K. Weiner, and A. J. Wilkinson, volume 2, pages 3–36 (ESCOM, Leiden, 1993).

[37] D. C. Rapaport, *The art of molecular dynamics simulation* (Cambridge University Press, Cambridge, 1995).

[38] D. Frenkel and B. Smit, *Understanding molecular simulation* (Academic Press, San Diego, 2002).

[39] H. Hu and W. Yang, Free energies of chemical reactions in solution and in enzymes with ab initio quantum mechanics/molecular mechanics methods. *Annu. Rev. of Phys. Chem.* 59, 573 (2008).

[40] H. M. Senn and W. Theil, QM/MM methods for biological systems, in *Atomistic approaches in modern biology: From quantum chemistry to molecular simulations*, Topics in Current Chemistry, Vol. 268, edited by M. Reiher, pp. 173–290 (Springer, Berlin, 2007).

[41] P. Attard, Simulation results for a fluid with the Axilrod-Teller triple dipole potential. *Phys. Rev. A* 45, 5649 (1992).

[42] A. Heyden, H. Lin, and D. G. Truhlar, Adaptive partitioning in combined quantum mechanical and molecular mechanical calculations of potential energy functions for multiscale simulations, *J. Phys. Chem. B* 111, 2231 (2007).

[43] J. Gao and C. Alhambra, A hybrid semiempirical quantum mechanical and lattice-sum method for electrostatic interactions in fluid simulations. *J. Chem. Phys.* 107, 1212 (1997).

[44] D. Riccardi, P. Schaefer, and Q. Cui, pKa calculations in solution and proteins with QM/MM free energy perturbation simulations: A quantitative test of QM/MM protocols. *J. Phys. Chem. B* 109, 17715 (2005).

[45] K. Nam, J. Gao, and D. M. York, An efficient linear-scaling ewald method for long-range electrostatic interactions in combined QM/MM calculations. *J. Chem. Theory Comp.* 1, 2 (2005).

[46] R. C. Walker, M. F. Crowley, and D. A. Case, The implementation of a fast and accurate QM/MM potential method in Amber. *J. Comp. Chem.* 29, 1019 (2008).

[47] T. Laino, F. Mohamed, A. Laio, and M. Parrinello, An efficient linear-scaling electrostatic coupling for treating periodic boundary conditions in QM/MM simulations. *J. Chem. Theory Comp.* 2, 1370 (2006).

[48] K. N. Kudin and G. Scuseria, A fast multipole algorithm for the efficient treatment of the Coulomb problem in electronic structure calculations of periodic systems with Gaussian orbitals. *Chem. Phys. Lett.* 289, 611 (1998).

[49] K. N. Kudin and G. E. Scuseria, Analytic stress tensor with the periodic fast multipole method. *Phys. Rev. B* 61, 5141 (2000).

[50] W. Im, S. Bernéche, and B. Roux, Generalized solvent boundary potential for computer simulations. *J. Chem. Phys.* 114, 2924 (2001).

[51] B. A. Gregersen and D. M. York, Variational electrostatic projection (VEP) methods for efficient modeling of the macromolecular electrostatic and solvation environment in activated dynamics simulations. *J. Phys. Chem. B* 109, 536 (2005).

[52] P. Schaefer, D. Riccardi, and Q. Cui, Reliable treatment of electrostatics in combined QM/MM simulation of macromolecules. *J. Chem. Phys.* 123, 014905/1 (2005).

[53] C. Tuma and H. Sauer, A hybrid MP2/planewave-DFT scheme for large chemical systems: proton jumps in zeolites. *Chem. Phys. Lett.* 387, 388 (2004).

[54] S. Svelle, C. Tuma, X. Rozanska, T. Kerber, and J. Sauer, Quantum chemical modeling of zeolite-catalyzed methylation reactions: toward chemical accuracy for barriers. *J. Amer. Chem. Soc.* 131, 816 (2009).

[55] R. A. Christie and K. D. Jordan, N-body decomposition approach to the calculation of interaction energies of water clusters, in *Intermolecular forces and clusters II*, Structure and Bonding series, series edited by D. M. P. Mingos, Volume 116, pages 27–41 (Springer, Berlin, 2005).

[56] E. Dahlke, H. Lin, H. Leverentz, and D. G. Truhlar, MBPAC-2009, 2009, University of Minnesota, Minneapolis, URL: http://comp.chem.umn.edu/mbpac/.

chapter six

Electron correlation in solids: Delocalized and localized orbital approaches

*So Hirata, Olaseni Sode, Murat Keçeli,
and Tomomi Shimazaki*

Contents

6.1 Introduction

There are two computational approaches to the electronic structures of periodic and nonperiodic solids—the delocalized and localized orbital approaches [98]. The delocalized orbital approach or the so-called *crystalline orbital* (CO) *theory* [3–5,18,24,63,72,107,123] is strongly based on the periodic boundary conditions. It views an extended system of one-dimensional periodicity as a ring of K identical repeat units. A symmetry-adapted orbital such as a canonical Hartree–Fock (HF) orbital in this approach is delocalized over the entire ring and has the Bloch form [14]:

$$\psi_{pk}(\mathbf{r}) = K^{-1/2} \sum_{\mu} \sum_{m} C_{pk}^{\mu} \exp\left(imka\right) \chi_{\mu}\left(\mathbf{r} - m\mathbf{a}\right), \quad (6.1)$$

where C_{pk}^{μ} is a CO coefficient, $\mathbf{a}$ is the lattice vector that outlines the unit cell, and $\chi_{\mu}\left(\mathbf{r} - m\mathbf{a}\right)$ is the μth atomic orbital (AO) centered in the mth unit cell. Each orbital is characterized by k, which is the linear momentum (in atomic units) of an electron in this orbital, and can take one of K distinct values:

$$\frac{ka}{\pi} = \frac{2m}{K}, \forall m = 1, 2, \ldots, K. \tag{6.2}$$

In the delocalized orbital approach, K is a crucial parameter with dual meaning. It is the number of k-vector sampling points in the first Brillouin zone (BZ) according to Equation (6.2). It is also the nominal size of the system, which should thus be commensurate with the number of the nearest neighbor unit cells included in the lattice sums of particle–particle interactions.

Quantities that enter the formalisms of CO theory are characterized by how they scale with respect to this parameter K. For instance, a CO scales as $K^{-1/2}$, as indicated by Equation (6.1), which is an immediate result of the normalization of the orbital. An orbital energy and a two-electron integral in the CO basis [47] are written as

$$e_{pk_p}^{HF} = K^0 \sum_{\mu,\nu} \sum_{m=-S}^{+S} C_{pk_p}^{\mu*} C_{pk_p}^{\nu} \exp(imk_pa) f_{\nu(m)}^{\mu(0)}, \tag{6.3}$$

$$v_{rk_r sk_s}^{pk_p qk_q} = K^{-1} \sum_{\kappa,\lambda,\mu,\nu} \sum_{m_1=-S}^{+S} \sum_{m_2=-L}^{+L} \sum_{m_3=m_2-S}^{m_2+S} C_{pk_p}^{\kappa*} C_{rk_r}^{\lambda} C_{qk_q}^{\mu*} C_{sk_s}^{\nu}$$

$$\times \exp\{i(m_1k_r - m_2k_q + m_3k_s)a\} v_{\lambda(m_1)\nu(m_3)}^{\kappa(0)\mu(m_2)}, \tag{6.4}$$

and scale as K^0 and K^{-1}, respectively. Here, S and L are the numbers of the nearest neighbor unit cells that are included in the lattice sums of short- and long-range interactions, respectively [47]. The matrices **f** and **v** are Fock and two-electron integrals in the AO basis. These definitions are obtained straightforwardly by assuming the CO of Equation (6.1) and simplifying the expressions with the following relation,

$$\sum_m \exp\{im(k_p - k_q)a\} = K\delta_{pq}, \tag{6.5}$$

where δ_{pq} is Kronecker's delta.

The significance of clarifying the K scaling of these quantities lies in the fact that it provides a rigorous basis on which to determine whether a certain electronic structure theory is size-extensive or not. Take second-order Møller–Plesset (MP2) perturbation theory [9, 43, 53, 71, 74, 75, 108, 109, 114–120, 139] as an example. The MP2 correlation energy for the entire ring [43, 47] is obtained by evaluating

$$E^{MP2} = \sum_{i,j,a,b} \sum_{k_i,k_j,k_a} \frac{v_{ak_a bk_b}^{ik_i jk_j} (2v_{ak_a bk_b}^{ik_i jk_j} - v_{bk_b ak_a}^{ik_i jk_j})^*}{e_{ik_i}^{HF} + e_{jk_j}^{HF} - e_{ak_a}^{HF} - e_{bk_b}^{HF}}. \tag{6.6}$$

According to Equations (6.3) and (6.4), the numerator and denominator scale as K^{-2} and K^0, respectively. The threefold k summation (the BZ integration) gives rise to a factor of K^3. Together, we find that E^{MP2} depends linearly on K and is thus size-extensive. The MP2 correlation energy per unit cell is E^{MP2}/K and is asymptotically independent of size (size-intensive), as it should be. Size-extensivity or the lack of this property in any other method can be judged in an analogous fashion. This criterion of size-extensivity based on the K scaling is equivalent to the usual diagrammatic criterion [11, 77].

In the localized orbital approach, the local energy E_i and the local wave function Ψ_i of subsystem i of a macromolecule are defined and evaluated by using orbitals that are spatially localized around this subsystem. The total energy and wave function are by definition the sum and product of these local quantities:

$$E = \sum_i^K E_i, \tag{6.7}$$

$$\Psi = \prod_i^K \Psi_i. \tag{6.8}$$

They are, therefore, size-extensive by construction regardless of how the local energies and wave functions are obtained. For instance, even a truncated configuration-interaction (CI) method, which is not size-extensive in the diagrammatic sense, can provide size-extensive total energies and wave functions in a localized orbital basis. Furthermore, the cost of evaluating the local energy and wave function is evidently independent of K in the $K \to \infty$ limit. Hence, the cost of evaluating the total energy E scales only linearly with respect to the number of subsystems, again, regardless of the details of the electronic structure theory used to determine E_i and Ψ_i.

We can view the approximation embodied by Equations (6.7) and (6.8), the lowest-rank truncation of the converging many-body expansions,

$$E = \sum_i E_i + \sum_{i<j}(E_{ij} - E_i - E_j)$$

$$+ \sum_{i<j<k}(E_{ijk} - E_{ij} - E_{jk} - E_{ki} + E_i + E_j + E_k) + \dots, \tag{6.9}$$

$$\Psi = \prod_i \Psi_i \cdot \prod_{i<j} \Psi_{ij}(\Psi_i \Psi_j)^{-1} \cdot \prod_{i<j<k} \Psi_{ijk}(\Psi_{ij}\Psi_{jk}\Psi_{ki})^{-1}\Psi_i\Psi_j\Psi_k \cdot \dots, \tag{6.10}$$

where E_{ij} and Ψ_{ij} are the energy and wave function of the dimer of subsystems i and j and E_{ijk} and Ψ_{ijk} the corresponding quantities of the trimer. Since the Hamiltonian in chemistry consists of one- and two-body interactions, we might expect that the truncation of these equations after the second term and the second factor is particularly useful. A finite-order truncation of these equations is size-extensive even when the electronic structure method to evaluate the local energies and wave functions is not diagrammatically size-extensive. However, as we include higher-order terms (that involve larger subsystems) more of the errors arising from the lack of size-extensivity are reintroduced.

How do the delocalized and localized orbital approaches relate to each other? They are related by a basis transformation such as a unitary transformation if both sets of orbitals are orthonormal. For instance, localized, orthonormal Wannier orbitals are defined by a Fourier transform of delocalized, orthonormal Bloch orbitals:

$$\zeta_{pm}(\mathbf{r} - m\mathbf{a}) = K^{-1/2} \sum_{k} \exp(-imka)\psi_{pk}(\mathbf{r}). \tag{6.11}$$

The two-electron integral and MP2 excitation amplitude in the Wannier orbital basis are given by

$$v_{rm_rsm_s}^{p0qm_q} = K^{-2} \sum_{k_q,k_r,k_s} \exp\{-i(m_r k_r - m_q k_q + m_s k_s)a\} v_{r k_r s k_s}^{p k_p q k_q}, \tag{6.12}$$

$$t_{am_abm_b}^{i0jm_j} = K^{-2} \sum_{k_j,k_a,k_b} \frac{\exp\{-i(m_a k_a - m_j k_j + m_b k_b)a\} v_{ak_a bk_b}^{ik_i jk_j}}{e_{ik_i}^{HF} + e_{jk_j}^{HF} - e_{ak_a}^{HF} - e_{bk_b}^{HF}} \tag{6.13}$$

and are independent of K (note that the two-electron integrals scale as K^{-1}, orbital energies K^0, and the threefold k-summation K^3). Using these quantities, we can transform the MP2 correlation energy for the entire ring, Equation (6.6), into

$$E^{MP2} = K \sum_{i,j,a,b} \sum_{m_a=-S}^{+S} \sum_{m_j=-L}^{+L} \sum_{m_b=m_j-S}^{m_j+S} t_{am_abm_b}^{i0jm_j} \left(2v_{am_abm_b}^{i0jm_j} - v_{bm_bam_a}^{i0jm_j}\right)^*. \tag{6.14}$$

In this localized orbital basis, the MP2 energy is K multiples of the sum that does not depend on K and that is thus interpretable as a local energy. Equation (6.14) is hence an explicitly size-extensive equivalent of Equation (6.6).

The delocalized and localized orbital approaches are complementary to each other and one should use either approach depending on the system that is being dealt with [98]. The delocalized orbital approach is nearly inevitable when treating systems with delocalized wave functions such as conjugated polymers and metals. In such treatments, a diagrammatically size-extensive electronic structure method must be used. Since these delocalized orbitals have well-defined linear momenta, energy bands—a central concept of solid-state physics—can be immediately available from these calculations. On the other hand, the localized orbital approach is well suited to weakly bonded systems such as molecular clusters and molecular crystals. In this approach, a distance-based truncation of lattice sums of particle–particle interactions can be particularly transparently introduced, although this is essential in any type of extended system calculations (see below). One can argue that this approach is strongly size-extensive and strongly linear-scaling in computational cost because any electronic structure method becomes overall size-extensive and linear-scaling when combined with this approach. Furthermore, disorders and nonperiodic macromolecules can be as straightforwardly treated by this approach as perfectly periodic crystalline solids. However, energy bands are hard (albeit not impossible [93]) to extract, which may be viewed as a result of the uncertainty principles operating between position and momentum.

In the following, we briefly describe our contributions to the methodological developments in these two approaches, also surveying representative applications that illustrate the strengths and weaknesses of each.

6.2   *Delocalized orbital approach*

6.2.1   *Methods*

Electronic structure methods can be largely divided into two categories: molecular orbital (MO) theory and density-functional theory (DFT). DFT in its Kohn–Sham (KS) formulation [69, 88] is a mean-field theory with its approximation entirely encapsulated in the so-called *exchange-correlation functional*, whose associated molecular integrals and energies are evaluated numerically using a multicenter quadrature grid. In MO theory, in addition to the mean-field theory known as HF theory, we have several series of systematic approximations converging toward the exact basis-set solutions of the Schrödinger equations [51]. They include many-body or MP perturbation theory [10], coupled-cluster (CC) theory [11], and CI theory [99]. A truncated CI is not diagrammatically size-extensive and is not considered further in this chapter with the exception of CI singles (CIS), which is size-intensive for excitation energies. The MO and DFT methods provide primarily energies and wave functions of molecules in the ground electronic states. The derivatives of energy with respect to atomic coordinates can also be evaluated analytically by virtue of the mathematical properties of Gaussian basis functions [95]. These derivatives are related to atomic forces and force constants and are invaluable in determining the equilibrium structures and vibrational frequencies. Furthermore, the energies and wave functions of electronic excited states can be obtained by the methods systematically derivable by applying the time-dependent linear response theory [38] to each of these MO and DFT approximations.

We made the following contributions to the CO methods [3–5, 24] within both the DFT and MO frameworks. We implemented a Gaussian-basis-set DFT method for extended systems of one- [49] and two-dimensional periodicity [129] (the pioneering work is due to Mintmire et al. [81,82]). The one-dimensional systems include helical polymers, which are hard to treat with planewave basis functions. Local, gradient-corrected, and hybrid exchange-correlation functionals (see Suhai for the initial use of hybrid functionals [116]) can be employed in conjunction with the density-fitting approximation of Dunlap et al. [27] to the classical Coulomb (J) energy and integrals. The long-range lattice sum of classical Coulomb (J) interactions was carried through to an infinite distance by virtue of the multipole expansion of Delhalle et al. [25]. We reported an initial implementation of analytical energy gradients with respect to atomic coordinates and lattice constants [42,129]. We found that the formula for the lattice constant gradients contained a surface integral that could be interpreted as a quadrature-weight derivative that was nonvanishing even in the limit of an infinitely fine grid. We proposed a convenient algorithm to evaluate this term with the Gauss theorem converting a surface integral to a volume integral. We performed the first Gaussian-basis-set time-dependent DFT (TDDFT) calculation for excitation energies in polymers [41], exposing its pathological behavior. TDDFT failed to reproduce exciton binding and this was traced to the incomplete cancellation of self-interaction that plagued almost all exchange-correlation functionals [38, 54]. We also introduced the Tamm–Dancoff approximation to TDDFT [40], which was applicable to atoms, molecules, and polymers alike, although it shared the same pathological behavior in polymers as TDDFT.

We reported a Gaussian-basis-set HF method for extended systems of one- [49] and two-dimensional periodicity [129], including the capability to evaluate energy gradients analytically [42] (analytical gradients were pioneered by Teramae et al. [127, 128]). See also Jacquemin and Champagne [56] for analytical gradients with respect to the helical angle and others [55, 59, 60, 70, 94, 110] for elastic modulus and stress tensors. We also introduced analytical second derivatives of energy with respect to atomic coordinates for polymers [44] (see also Sun and Bartlett [121]), implementing a periodic coupled-perturbed HF (CPHF) equation solver [19, 64, 65]. Electronic excited states in polymers were probed by CIS and time-dependent HF (TDHF) [41], which had been considered earlier by Suhai [107, 110–113] and Vračko et al. [135–137]. We also explored electron correlation in polymers at the MP2 [43] and MP3 levels [39, 47] as well as at the CC singles and doubles (CCSD) level and its various approximations [39, 47]. See also Förner et al. [28, 29], Ye et al. [139], and Reinhardt [96] for other CC implementations on polymers. We reported analytical gradients of MP2 energies [43] and proposed MP2 for electronic excited states [37]. The latter was based on quasi-particle (i.e., correlated) energy bands of polymers obtained by the many-body Green's function theory (MBGF) at the levels of MP2 and the second-order approximation to the inverse Dyson self-energy [91,92,109,118].

6.2.2   Applications

Figure 6.1 compares the calculated phonon dispersion curves of polyethylene [45] with two sets of experimentally derived curves [104,124–126]. The calculation was based on the DFT CO method with the Slater–Vosko–Wilk–Nusair (SVWN) functional and the 6-31G* basis set. The interaction force constant matrices necessary in computing the dispersion were obtained by a supercell approach employing C_7H_{14} as a repeat unit. DFT is known to work particularly well for vibrational frequencies of organic molecules and the overall agreement between theory and experiments attests to the fact that this is also the case with polyethylene. However, quantitative agreement seems too good as the mean square error of 21 cm^{-1} is achieved for the infrared- and Raman-active ($k = 0$) phonons without the inclusion of the effects of vibrational anharmonicity. This accurate agreement

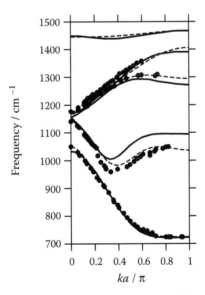

Figure 6.1 Phonon dispersion curves of polyethylene. The solid curves are computed by the SVWN/6-31G* method [45] and the broken curves by an empirical force field of Tasumi et al. [124–126]. The dots represent the observed frequencies of *n*-alkanes (Snyder and Schachtschneider [104]).

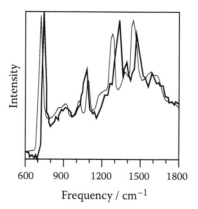

Figure 6.2 Inelastic neutron scattering spectra of polyethylene. The thin curves are computed by the SVWN/6-31G* method [45] and the solid curves are measured by Parker [87].

is, therefore, fortuitous and is expected to deteriorate when anharmonicity can be included in the calculations [62].

Figure 6.2 is a comparison between simulated [45] and observed [87] spectra of inelastic neutron scattering from polyethylene. The spectra probe the density of vibrational states that involve hydrogen motion in particular. The simulated spectrum based on the phonon dispersion curves shown in Figure 6.1 reproduces the observed spectrum remarkably well. In fact, without a pilot of quantitative simulation such as this, it is hard to interpret inelastic neutron scattering spectra of polymers because of their generally low resolution and congestion of peaks, some of which are due to overtones and combinations. Again, while there is no question about the usefulness of the DFT calculation for this purpose, the accurate agreement in peak positions between theory and experiment is to some extent fortuitous.

The next example is photoelectron spectra of *trans*- and *cis*-poly-acetylenes (Figure 6.3). The calculated spectra were obtained by the DFT CO method with the Becke3–Lee–Yang–Parr (B3LYP) hybrid functional and the 6-31G* basis set [49]. Each peak in the electronic densities of states was convoluted with an asymmetric band shape. Agreement between theory and experiment [57] is remarkable both in peak positions and intensity profiles. The calculations can apparently reproduce the slight difference in the appearance of the spectra between the two isomers. However, the accurate agreement in peak positions must be viewed with caution. It is well known that DFT with a semilocal exchange-correlation functional, namely without admixture of the HF exchange, fails to describe the electronic and geometrical structures of *trans*-polyacetylene qualitatively correctly [6, 49, 83, 86, 116, 133, 134]. It predicts a metallic electronic

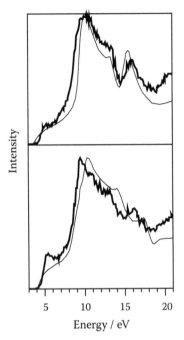

Figure 6.3 Photoelectron spectra of *trans-* (top) and *cis*-polyacetylenes (bottom). The thin curves are computed by the B3LYP/6-31G* method [49] and the solid curves are measured by Kamiya et al. [57].

structure with an equidistant equilibrium geometry in a stark violation of Peierls theorem. In general, semilocal exchange-correlation functionals tend to underestimate the ionization energies and fundamental band gaps of solids. In contrast, HF theory overestimates (see below) these quantities. Hence, the accurate agreement in peak positions (if not intensity profiles) obtained by the B3LYP functional is largely due to cancellation of errors between the semilocal DFT functionals and HF exchange that enter this hybrid, which have large errors with opposite signs [48].

All of the aforementioned examples are based on DFT and, while useful for interpretive purposes, are not considered predictive. Can we apply converging series of approximations such as those based on MP and CC theories directly to solids? The answer is, in principle, affirmative because they are size-extensive methods, but in practice, it has been nearly negative because of the immense computational costs of these calculations. For instance, the MP2 energy expression, Equation (6.6), involves a three-fold k-summation, making the cost of its evaluation proportional to K^3 asymptotically (note that the integral transformation step is usually even more time consuming). Likewise, the cost of evaluating MP3 and CCSD energies scale as K^4 [47]. To obtain well-converged HF energies and wave functions, one typically needs to set $K = 10$ or even $K = 20$, rendering K^3 and K^4 very large and the MP and CC calculations unfeasible in a realistic time frame.

Let us seek a way to drastically reduce the computational costs of correlated calculations. The key is to recognize that the correlation interactions summed in MP and CC decay much more rapidly than the slowest-decaying interactions in a HF energy [36]. Equation (6.12) indicates that the two-electron integrals that enter the MP2 energy expression, Equation (6.14), represents an electrostatic interaction between two local charge distributions whose net charges are zero. At a large distance (r), the interaction reduces to a dipole–dipole interaction that decays as r^{-3}. According to Equation (6.13), double excitation amplitudes in a local basis must also decay as r^{-3} because they are related to the two-electron integrals of Equation (6.12). Together, we find that the correlation interactions in MP2 exhibit r^{-6} asymptotic behavior [8,47] and are viewed essentially as dispersion effects. This is true in any solids insofar as the perturbation treatment of correlation is convergent (see, however, [122]). Since the unit cell of a solid must be neutral [25], the classical Coulomb (J) interactions in an HF energy are of the dipole–dipole type in the long range and decay as r^{-3} [47]. The exchange interactions (K) are known to decay exponentially in insulators since the density matrix elements decay exponentially with the exponents being proportional to the fundamental band gaps [26,84,90]. Figure 6.4 illustrates that excitation amplitudes and density matrix elements in the atomic orbital basis display these expected asymptotic decay behaviors as early as in the third nearest neighbor cells [47]. In these and more distant cells, the correlation interactions are much smaller than the slowest-decaying, classical Coulomb (J) interactions summed in an HF energy.

This suggests that two separate truncation radii for lattice sums for HF (a large radius) and correlation (a small radius) energies can be used [48,100]. Accordingly, a much smaller value of K (as a measure of system size) can be adopted only at the correlation step, while keeping the value of K sufficiently large at the HF step. In other words, it should be justified to downsample k-points in the BZ integrations only in the correlation step of the calculations, which can lead to dramatic savings in operational and storage costs. We examined the effect of this approximation on the MP2 correlation energy per unit cell:

$$E_{cell}^{MP2} = K^{-1}n^3 \sum_{i,j,a,b} \sum_{k_i,k_j,k_a \in K_n} \frac{v_{ak_a bk_b}^{ik_i jk_j} \left(2v_{ak_a bk_b}^{ik_i jk_j} - v_{bk_b ak_a}^{ik_i jk_j}\right)^*}{e_{ik_i}^{HF} + e_{jk_j}^{HF} - e_{ak_a}^{HF} - e_{bk_b}^{HF}}, \qquad (6.15)$$

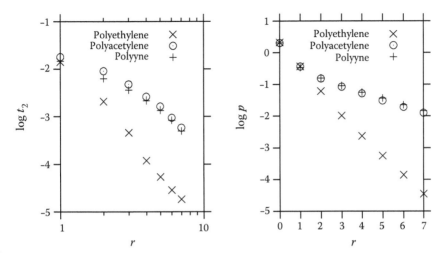

Figure 6.4 The largest double excitation (t_2) amplitudes in the AO basis of the CCSD wave functions for polyethylene, *trans*-polyacetylene, and polyyne plotted as a function of distance measured in the units of $|a|$ (left) [47]. The largest density matrix elements (p) in the AO basis of the HF wave functions plotted as a function of distance (right) [47].

where K_n is a set of k-vectors that satisfy the following condition, assuming that n evenly divides K,

$$\frac{ka}{\pi} = \frac{2\pi n m}{K}, \forall m = 1, 2, \ldots, K/n. \tag{6.16}$$

Table 6.1 shows that the MP2/6-31G* correlation energy of polyethylene can be recovered within 0.3% of the converged value by using every 5th (the mod-5 approximation) of the k-points used in the preceding HF calculation ($K = 20$). A 15-fold speedup of the MP2 calculation was observed. In other words, the number of the nearest neighbor C_2H_4 unit cells included in the lattice sums for the MP2 energy can be as small as two, while it is ten for the HF energy, without a noticeable loss in accuracy. Surprisingly, sampling just one k-point at the BZ center (Γ-point) can still yield 102.7% of the converged correlation energy at a small fraction of the usual computational cost. This Γ approximation has an additional benefit of making any electron-correlation method size-extensive because no quantities in the formalism depend on K. These mod-n approximations are even more effective in MP3 and CCSD, where we observed a speedup by nearly three orders of magnitude with reasonably accurate correlation energies [61].

Table 6.1 The Effects of Downsampling in the BZ
Integrations for the MP2/6-31G* Calculations of
Polyethylene [100]. The HF/6-31G* Calculation was
Carried Out with $K = 20$

Approximation	K	Error in E_{cell}^{MP2} (%)	Speedup
Exact	20	0	1
mod 2	10	0.1	3.8
mod 4	5	0.3	11
mod 5	4	0.3	15
Γ (mod 20)	1	2.7	81

The question is whether the observed accuracy in the *total* correlation energies is high enough for *relative* energies. The many-body Green's function (MBGF) method in the MP2 approximation provides a way to compute correlated ionization and electron-attachment energies (the quasi-particle energy bands) by evaluating the formula similar to the MP2 energy expression [37, 91, 92, 109, 118]. In the mod-n approximation, the formula reads

$$
e_{pk_p}^{MP2} = e_{pk_p}^{HF} + n^2 \sum_{j,a,b} \sum_{k_j,k_b \in K_n} \frac{v_{ak_a bk_b}^{pk_p jk_j} \left(2v_{ak_a bk_b}^{pk_p jk_j} - v_{bk_b ak_a}^{pk_p jk_j} \right)^*}{e_{pk_p}^{HF} + e_{jk_j}^{HF} - e_{ak_a}^{HF} - e_{bk_b}^{HF}}
$$

$$
+ n^2 \sum_{i,j,b} \sum_{k_j,k_b \in K_n} \frac{v_{pk_p bk_b}^{ik_i jk_j} \left(2v_{pk_p bk_b}^{ik_i jk_j} - v_{bk_b pk_p}^{ik_i jk_j} \right)^*}{e_{pk_p}^{HF} + e_{bk_b}^{HF} - e_{ik_i}^{HF} - e_{jk_j}^{HF}}. \tag{6.17}
$$

Figure 6.5 shows the HF/6-31G* valence energy bands (broken curves) of polyethylene obtained with $K = 120$ [48]. The large value of K is unnecessary for obtaining a converged HF energy, but it is desirable for smooth energy bands and reliable density of states and can be easily used in an HF calculation because its cost is almost independent of K. However, such a large K value would make the subsequent MP and CC calculations unfeasible unless an aggressive downsampling is adopted by the mod-n approximation. For polyethylene, a downsampling by a factor of as large as 20 with only six k-vectors in the BZ in the MP2 step seems justified, yielding the quasi-particle (correlated) energies (open circles) at a sparse grid of these k-vectors. Continuous quasi-particle energy bands can be obtained by interpolation, combining the smooth HF energy bands and sparse MP2 quasi-particle energies with the following formula:

$$
e_{pk}^{MP2} = e_{pk}^{HF} + \left(e_{pk_1}^{MP2} - e_{pk_1}^{HF} \right) \frac{k - k_2}{k_1 - k_2} + \left(e_{pk_2}^{MP2} - e_{pk_2}^{HF} \right) \frac{k - k_1}{k_2 - k_1}, \tag{6.18}
$$

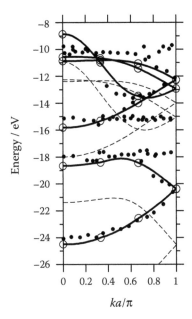

Figure 6.5 Valence energy bands of polyethylene obtained by the HF/6-31G* method (broken curves) with $K = 120$ and the MP2/6-31G* method (open circles) with $K = 6$ [61]. The solid curves are obtained by interpolation of the MP2/6-31G* quasi-particle energies. The dots are the peak positions of angle-resolved photoelectron spectra of n-$C_{44}H_{90}$ reported by Yoshimura et al. [140].

where k_1 and k_2 are the two nearest k-vectors at which MP2 quasi-particle energies are explicitly evaluated in the mod-n approximation. The solid curves in Figure 6.5 are the MP2 energy bands obtained by interpolation [48] and they agree accurately with the bands observed by angle-resolved photoelectron spectroscopy [140]. The HF bands, in contrast, overestimate both the ionization energies and band widths considerably. The mod-n approximation, therefore, works well for relative energies as computed by Equation (6.17).

The valence energy bands of *trans*- and *cis*-polyacetylenes obtained by the HF and mod-n MP2 methods [48] are compared in Figure 6.6. The results of the *trans* isomer indicate that the correlation corrections to energy bands are substantial (a few eV) and that the interpolated mod-n MP2 energy bands with $K = 6$ recover the vast majority of these corrections and agree well with the MP2 energy bands with $K = 24$. The latter calculation took 12 times as much CPU time as the former, while the former cost less than 2

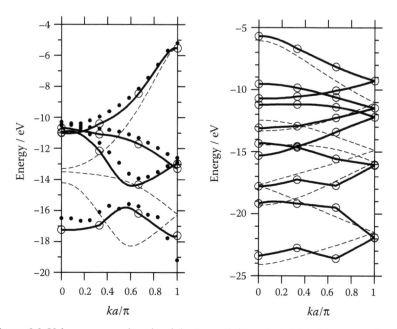

Figure 6.6 Valence energy bands of the *trans* (left) and *cis* (right) isomers of poly-acetylenes [48]. The broken curves are obtained by the HF/6-31G* method with $K = 120$, the open circles by the MP2/6-31G* method with $K = 6$, and the solid curves by interpolation of the MP2/6-31G* quasi-particle energies. The dots are the MP2/6-31G* quasi-particle energies obtained with $K = 24$.

CPU hours on a 3.2-GHz Intel Xeon processor. The interpolated MP2 energy bands of the *cis* isomer suggest that the bands that lie below −17 eV are nonsmooth and suspected of some larger errors than those that lie above the threshold.

Figure 6.7 compares the HF and mod-n MP2 densities of states with photoelectron spectra of *trans-* and *cis*-polyacetylenes [57]. In both isomers, the HF calculations place peaks too high in energies, which may be expected from the tendency of HF theory to overestimate ionization energies. In the *cis* isomer, the HF density of states has significant difference in appearance from the experiment and the correct interpretation of the latter seems hopeless. The MP2 densities of states, in contrast, achieve an equal and respectable degree of agreement with the observed spectra. Unlike the B3LYP results [49] discussed earlier, the good agreement in peak positions as well as in intensity profiles between the MP2 and experimental results is due much less to cancellation of errors.

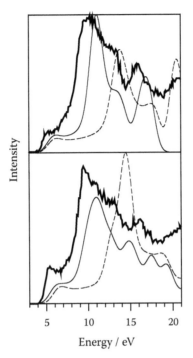

Figure 6.7 Photoelectron spectra of *trans*- (top) and *cis*-polyacetylenes (bottom). The thin solid curves are computed by the MP2/6-31G* method [48], the thin broken curves by the HF/6-31G* method [48], and the solid curves measured by Kamiya et al. [57].

6.3 Localized orbital approach

6.3.1 Methods

For a large cluster or a crystal that consists of relatively weakly interacting molecules, the MOs of these molecules can serve as a convenient localized orbital basis on which to decompose the total energy into its many-body components:

$$E = \sum_i E'_i + \sum_{i<j}(E'_{ij} - E'_i - E'_j) + \ldots . \tag{6.19}$$

When the energies of the monomer, dimer, and so forth are assessed in a vacuum, however, this series is not rapidly convergent and the truncation after the dimer sum constitutes a rather poor approximation. Instead, we proposed [35, 50, 58] evaluating the correlated energies of monomers and dimers in the presence of the rest of the cluster or crystal constituents represented by atomic partial charges or molecular dipole moments determined self-consistently at the HF level. The primes in Equation (6.19) denote this inclusion of electrostatic environmental effects. Therefore, even an approximation that truncates Equation (6.19) after the dimer sum includes three-body and higher-order many-body effects that are in the category of classical Coulomb (J) interactions; it includes the so-called induction or (hydrogen-bond) cooperativity in the dipole–dipole or point-charge approximation at the HF level, while accounting for the one- and two-body kinetic, exchange, and correlation (hence dispersion) effects at a correlated level.

Many groups are working on similar, easily implemented methods suitable for molecular clusters and crystals as well as for a certain class of covalently bonded macromolecules. They include Gadre et al.'s molecular tailoring approach (MTA) [32], Kitaura et al.'s pair-interaction [67] and fragment molecular orbital (FMO) methods [66], Gordon et al.'s effective fragment potential (EFP) method [33], the incremental scheme of Fulde, Stoll, Paulus, and coworkers [30, 89, 106] (see Chapter 3), Zhang and Zhang's molecular fragmentation with conjugated caps (MFCC) [142], Deev and Collins' systematic fragmentation method (SFM) [20, 23], Li et al.'s cluster-in-molecule (CIM) method [73], Tschumper's multicentered (MC) integrated QM:QM approach [130], the electrostatically embedded many-body expansion (EE-MB) of Dahlke and Truhlar [21, 22], and the many-body expansion with a polarizable force field of Beran [12]. Manby et al. [76] applied an incremental scheme to an accurate determination of cohesive energies of an ionic solid. Ringer and Sherrill [97] reported a benchmark cohesive energy of crystalline benzene in an analogous scheme. Our method [35,50,58] is a direct simplification of Kitaura's pair-interaction method [67] that can be much more easily implemented than the original scheme without impairing its intrinsic high accuracy. See also the many-body scheme of O'Neill et al. in Chapter 7 of this volume.

Our methods [35, 50, 58] are called the binary- or ternary-interaction methods depending on the truncation of Equation (6.19). Their advantages are as follows: (1) They can achieve the accuracy of a few kcal/mol in energy differences between isomers of a medium-sized water cluster [50]. (2) They form a systematically converging series of approximations that are characterized by the truncation of Equation (6.19). (3) They are fast in the sense that the cost of a binary-interaction calculation scales quadratically even for the smallest clusters and linearly for large clusters and crystals [50]. The scaling can be sublinear for size-intensive quantities such as excitation energies [50]. (4) They are strongly size-extensive and can be combined with any electron-correlation method to give meaningful results. (5) They can be implemented extremely easily given a well-developed electronic structure program for molecules. (6) The function counterpoise corrections for basis-set superposition errors (BSSE) can be made conveniently during the evaluation of the dimer energies [58]. (7) Geometrical derivatives of energy can be computed at a linear-scaling cost regardless of the rank of derivatives or the details of derivative algorithms (numerical versus analytical). (8) Nonperiodic macromolecules and disorders in solids are treatable as easily as perfectly periodic solids.

The disadvantages of the binary- and ternary-interaction methods are the following: (1) Each of the monomers, dimers, and so forth must be unambiguously assignable an integer number of electrons for the calculation to be executable. (2) The many-body expansion of Equation (6.19) must be rapidly convergent for the result of the calculation to be meaningful. (3) In solid-state applications, there is no obvious way of extracting energy bands (i.e., the linear momenta of electrons). Because of (1) and (2), the method is not applicable to anionic water clusters, clusters of lithium atoms, or diffuse excited states, for instance.

In this chapter, we will concentrate on the solid-state application of the binary-interaction method [35], although it has also been applied to molecular clusters [50,58]. For an extended system of one-dimensional periodicity, the expression for the energy per unit cell in this approximation is

$$E_{cell} = \sum_i E'_{i(0)} + \sum_{i<j} \{E'_{i(0)j(0)} - E'_{i(0)} - E'_{j(0)}\}$$

$$+ \frac{1}{2} \sum_{m=-L}^{+L} (1 - \delta_{m0}) \sum_{i,j} \{E'_{i(0)j(m)} - E'_{i(0)} - E'_{j(m)}\}, \qquad (6.20)$$

where "$j(m)$" denotes the jth monomer in the mth unit cell and thus $E'_{i(0)j(m)}$ is an energy of the dimer that consists of ith monomer in the central unit cell and jth monomer in the mth unit cell. The primes indicate that the monomers and dimers are embedded in the electrostatic field created by a (large but finite) lattice of self-consistently determined atomic partial charges or dipole moments.

Energy gradients are obtained by evaluating

$$\frac{\partial E_{cell}}{\partial x} = \sum_i \frac{\partial E'_{i(0)}}{\partial x} + \sum_{i<j} \left\{ \frac{\partial E'_{i(0)j(0)}}{\partial x} - \frac{\partial E'_{i(0)}}{\partial x} - \frac{\partial E'_{j(0)}}{\partial x} \right\}$$

$$+ \frac{1}{2} \sum_{m=-L}^{+L} (1 - \delta_{m0}) \sum_{i,j} \left\{ \frac{\partial E'_{i(0)j(m)}}{\partial x} - \frac{\partial E'_{i(0)}}{\partial x} - \frac{\partial E'_{j(m)}}{\partial x} \right\}. \qquad (6.21)$$

Strictly speaking, this is an approximate formula because the derivatives of the embedding field with respect to an atomic coordinate (x) have not been taken. The contributions from such derivatives have been found negligible numerically insofar as the truncation radius of the lattice sum (L) is sufficiently large [35]. However, for the gradient with respect to the lattice constant (a), this is not the case even for a large value of L and a correction term accounting for a long-range electrostatic contribution is essential [35]. The expression for the gradient with respect to the lattice constant along the z-axis then reads

$$\frac{\partial E_{\text{cell}}}{\partial a} = \frac{1}{2} \sum_{m=-L}^{+L} \sum_{i,j} \sum_{\gamma} m \left\{ \frac{\partial E'_{i(0)j(m)}}{\partial z_{\gamma}^{j(m)}} - \frac{\partial E'_{j(m)}}{\partial z_{\gamma}^{j(m)}} \right\} + \frac{\partial E_{\text{LR}}}{\partial a}, \tag{6.22}$$

where $z_{\gamma}^{j(m)}$ is the z-coordinate of atom γ in the mth unit cell and E_{LR} is the sum of the dipole–dipole interaction energies between the central and distant unit cells:

$$E_{\text{LR}} = \frac{1}{2} \sum_{|m|>L} \frac{\mathbf{d} \cdot \mathbf{d} - 3(\mathbf{d} \cdot \hat{\mathbf{r}}_m)^2}{|\mathbf{r}_m|^3}, \tag{6.23}$$

where $\mathbf{d}$ is the unit cell dipole and $\mathbf{r}_m = m\mathbf{a}$.

Hessian and higher-order derivatives can be obtained by evaluating expressions that are analogous to Equation (6.21). Since our method is only loosely tied to the periodic boundary conditions in the sense that the underlying orbitals are localized and not symmetry adapted, the differentiating variables (x) do not have to be in-phase atomic coordinates that preserve the periodicity of the system. They can be individual atomic coordinates, leading to a complete set of interaction force constants and phonon dispersion curves in the entire BZ. Note that obtaining the whole phonon dispersion curves is not straightforward with the CO methods because atomic displacements along a normal mode of $k \neq 0$ phonon lift periodicity that underlies the CO formalisms (for instance, a supercell approach was used to obtain Figure 6.1 [45]).

6.3.2 Applications

Solid formic acid is highly anisotropic and can be computationally modeled as a one-dimensional hydrogen-bonded chain; the interchain interactions, which are electrostatic, are expected to be much weaker than the intrachain interactions [15, 132]. There are at least three plausible conformations of the one-dimensional chain: α, β_1, and β_2 (Figure 6.8). There is a surprising level of confusion and controversy surrounding the structure and spectra of solid formic acid for such a fundamental and simple crystalline solid. The controversy began with the report of infrared spectra of solid formic acid by Millikan and Pitzer [80]. They observed that many of the infrared bands of the solid were split into doublets and they assigned the doublets to in-phase and out-of-phase vibrations of adjacent formic acid

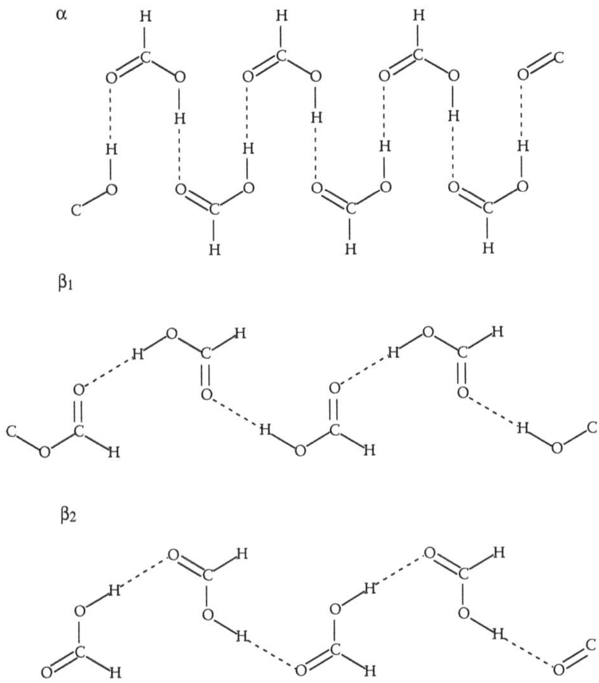

Figure 6.8 Structures of the α, β_1, and β_2 conformers of the one-dimensional formic acid chain.

molecules. Mikawa et al. [79], on the other hand, interpreted the splitting to be due to the coexistence of the α and β_1 structures, i.e., polymorphism. Zelsmann et al. [141] proposed that there was a polymorphic phase involving β_1 and β_2 above a certain transition temperature. In the meantime, diffraction studies [1,52,85] established unequivocally that the dominant structure of solid formic acid was β_1. The question is, therefore, whether there is a secondary structure in solid formic acid and, if so, whether it is α or β_2.

We optimized the structures of the three conformers at the MP2/aug-cc-pVDZ level using the binary-interaction method with atomic partial charges as the embedding potential [35]. We also carried out single-point energy calculations at the MP2/aug-cc-pVTZ and CCSD/aug-cc-pVDZ levels including BSSE corrections to the former. Both calculations found the β_1 conformer to be most stable, in accordance with the diffraction studies, followed by the β_2 conformer by 0.3 kcal/mol. The α conformer was least stable by 1.3–1.5 kcal/mol. From these energetics and also from the fact

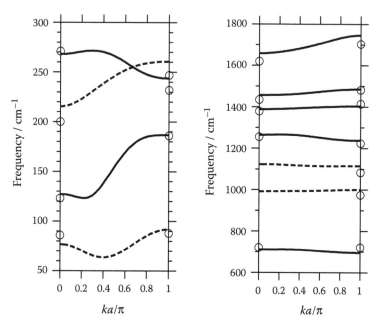

Figure 6.9 Phonon dispersion curves of solid formic acid (β_1) computed by the MP2/aug-cc-pVDZ method [35]. The binary-interaction method with atomic partial charges and the harmonic approximation are employed. The solid curves are in-plane phonons and the broken curves are out-of-plane phonons. The open circles are the positions of the observed infrared or Raman bands [15,31,78,80].

that the α conformer had significantly different lattice constants than the other two and the observed values, we concluded that the presence of the α conformer in solid formic acid was highly unlikely; we ruled out the hypothesis of Mikawa et al. [79] involving the α–β_1 polymorphism.

If there is no polymorphism, how can we explain the splitting of the infrared bands? To answer this, we computed the phonon dispersion curves of the three conformers at the MP2/aug-cc-pVDZ level in the harmonic approximation [35]. The calculated phonon dispersion curves of the β_1 conformer are shown in Figure 6.9. The open circles plot the positions of the observed infrared and Raman bands [15, 31, 78, 80] and they fall nicely on the left and right ends of the phonon dispersion curves where the phonons can be infrared and/or Raman active [132]. Therefore, there is no need to invoke polymorphism to explain the splitting of the infrared bands. A doublet can be assigned, without difficulty, to the in-phase and out-of-phase vibrations of adjacent formic acid molecules of the β_1 form, as Millikan and Pitzer originally suggested [80]. The splitting of the infrared bands need not imply polymorphism.

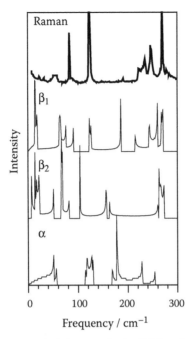

Figure 6.10 Raman spectrum (solid curves) of solid formic acid [34] and densities of states (thin curves) of the β_1, β_2, and α conformers computed by the MP2/ aug-cc-pVDZ method [35].

This is supported by a low-frequency Raman spectrum (Figure 6.10). We compared the observed spectrum [34] in the 0–300 cm^{-1} region with the phonon densities of states obtained at the MP2/aug-cc-pVDZ level for the α, β_1, and β_2 conformers [35]. Since the latter do not take into account the cross sections of Raman scattering, the intensities do not have to match, but the peak positions should agree between theory and experiment. The densities of states of the β_1 conformer are in good agreement with those observed, while the densities of the other two conformers do not reproduce many of the observed peak positions and have peaks where no bands are observed. Furthermore, there are no peaks in the observed spectrum that are uniquely assignable to a peak in the density of states of α or β_2. This is a piece of evidence suggesting that there was no secondary structure or polymorphism in the sample from which the Raman spectrum was measured.

Further evidence supporting this conclusion comes from a comparison between the simulated [35] and observed [13] inelastic neutron scattering spectra (Figure 6.11). Excellent agreement between theory and experiment can be seen when the β_1 conformation is assumed. The simulated spectrum

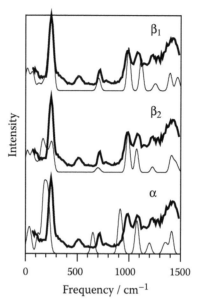

Figure 6.11 Inelastic neutron scattering spectra (solid curves) of solid formic acid [13] and hydrogen-amplitude weighted density of states (multiplied by a Debye–Waller factor) of the β_1, β_2, and α conformers computed by the MP2/aug-cc-pVDZ method [35].

of the β_2 conformer has difficulty reproducing the observed in the low-frequency region, whereas the calculated peak positions of the α conformer are shifted from those observed in the entire frequency domain considered. The calculations were performed for two other isotopomers, supporting the same conclusion. The simulated spectra do not include the effects of overtones and combinations, which give rise to intense peaks in inelastic neutron scattering spectra, and this is why the peak at around 500 cm^{-1} has not been reproduced by theory.

In summary, solid formic acid is likely composed of the most stable β_1 conformer and the observed infrared, Raman, and inelastic neutron scattering spectra can be most straightforwardly interpreted when the pristine β_1 structure is assumed. The splitting in the infrared bands can be assigned to the left and right ends of phonon dispersion curves that are infrared active and close in frequency. The available diffraction and spectroscopic data, in light of our correlated calculations, do not constitute evidence supporting polymorphism.

The next example is solid hydrogen fluoride, which can also be modeled as a one-dimensional hydrogen-bonded zigzag chain [131]. Earlier, we applied CO theory at the HF and B3LYP levels with the 6-311++G** basis set to an infinite chain of hydrogen fluoride [46]. We optimized its

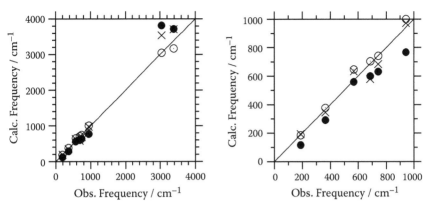

Figure 6.12 Frequencies of the infrared- and Raman-active phonons of solid hydrogen fluoride. The filled circles are the anharmonic frequencies obtained by the HF/aug-cc-pVDZ method [105], crosses are the harmonic frequencies obtained by the MP2/aug-cc-pVDZ method [105], and open circles are the anharmonic frequencies obtained by the combination of MP2 and CCSD(T) with the aug-cc-pVDZ basis set [105]. The observed frequencies are taken from [2,68].

geometry and calculated harmonic frequencies of the infrared- and Raman-active ($k = 0$) phonons. On the basis of the B3LYP calculations, which can adequately take account of the hydrogen-bond cooperativity, we proposed complete assignments of the infrared and Raman bands, which seem most trustworthy to this day. What could not be done then with the CO calculations were inclusions of electron correlation at the MP2 or higher-ranked methods, a correction of BSSE, a treatment of vibrational anharmonicity, and a calculation of phonon dispersion curves in the entire BZ.

Today, all of these can be easily accomplished with the binary-interaction method [105]. The BSSE-corrected MP2/aug-cc-pVTZ and CCSD/aug-cc-pVDZ methods could be used to determine the equilibrium geometry of an infinite chain of hydrogen fluoride and its harmonic force constants, normal modes, and frequencies. Furthermore, one- and/or two-dimensional slices of the potential energy surface [17] along the normal modes of $k = 0$ phonons were computed at the CCSD(T) and MP2 levels with the aug-cc-pVDZ basis set. These potentials were used to determine *anharmonic* frequencies of the infrared- and Raman-active phonons with the vibrational MP2 method [138]. They are compared with the observed frequencies [2,68] in Figure 6.12. With inclusion of either electron correlation or anharmonic effects, the errors in the calculated frequencies (filled circles and crosses) of the H–F stretching modes can be in excess of several hundred cm^{-1}; only when both electron correlation and anharmonicity are taken into account (open circles), can quantitative agreement be obtained. In the frequency region of the pseudo-translational and

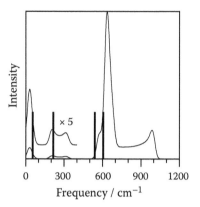

Figure 6.13 Observed peak positions (solid impulses) of the inelastic neutron scattering spectra of solid hydrogen fluoride [7,16] and hydrogen-amplitude weighted density of states (thin curves) computed by the MP2/aug-cc-pVDZ method [105].

librational modes (right panel), the most accurate agreement is between the CCSD(T)+MP2/aug-cc-pVDZ anharmonic calculations and the observed. It can be seen that the harmonic MP2 results suggest that band assignments for a pair of librational modes be swapped. When anharmonic effects are included, the same band assignments originally suggested by the B3LYP method [46] are recovered. The new calculation not only includes electron correlation and anharmonic effects more rigorously than the B3LYP calculation, but it does so in a systematic way that allows accuracy to be improved further if needed.

Unlike CO methods, the binary-interaction methods for molecular crystals are not strongly dependent on the periodic boundary conditions and can compute phonon dispersion curves in the entire BZ and phonon densities of states. Figure 6.13 shows the inelastic neutron scattering spectrum of solid hydrogen fluoride simulated on the basis of the phonon density of states obtained at the MP2/aug-cc-pVDZ level [105]. It explains the appearances of four peaks observed in a low-resolution inelastic neutron scattering spectra measured for this solid [7,16]. The predicted spectrum will be helpful in interpreting a high-resolution spectrum when it becomes available in the future.

6.4   Concluding remarks

Converging series of size-extensive, many-body methods for electrons and vibrations can potentially transform computational solid-state physics, surface science, and polymer and materials research by enabling computational characterizations of solids whose fidelity can be systematically improved. MP2, CCSD, and higher-ranked methods (not just for electrons but also for phonons) that go beyond usual DFT approximations are beginning to be feasible for solids or even liquids, thanks to algorithmic breakthroughs made in the past decades. These breakthroughs are ultimately based on our knowledge of different types of chemical interactions and their long-range behavior (and short-range behavior also [51,101–103]) and many-body properties. They also rely on our ability to sum those interactions most economically by truncating their many-body expansions expressed in optimized (localized, delocalized, explicitly correlated, etc.) bases. Significant research efforts are devoted to this field by both quantum chemists and solid-state physicists and are expected to intensify further in the near future.

Acknowledgments

This chapter is based on the lecture given by So Hirata at the 13th International Congress of Quantum Chemistry held in June 2009 at Helsinki. This work has been financially supported by the U.S. Department of Energy (DE-FG02-04ER15621), the U.S. National Science Foundation (CHE-0844448), and the Donors of the American Chemical Society Petroleum Research Fund (48440-AC6). So Hirata is a Camille Dreyfus Teacher-Scholar.

References

[1] A. Albinati, K. D. Rouse, and M. W. Thomas, Neutron powder diffraction analysis of hydrogen-bonded solids. II. Structural study of formic acid at 4.5 K, *Acta Cryst. B* 34, 2188 (1978).

[2] A. Anderson, B. H. Torrie, and W. S. Tse, Raman spectra of crystalline HF and DF, *Chem. Phys. Lett.* 70, 300 (1980).

[3] J. M. André, Self-consistent field theory for electronic structure of polymers, *J. Chem. Phys.* 50, 1536 (1969).

[4] J.-M. André, L. Gouverneur, and G. Leroy, L'étude théorique des systèmes périodiques. I. La méthode LCAO-HCO, *Int. J. Quantum Chem.* 1, 427 (1967).

[5] J.-M. André, L. Gouverneur, and G. Leroy, L'étude théorique des systèmes périodiques. II. La méthode LCAO-SCF-CO, *Int. J. Quantum Chem.* 1, 451 (1967).

[6] J. Ashkenazi, W. E. Pickett, H. Krakauer, C. S. Wang, B. M. Klein, and S. R. Chubb, Ground state of *trans*-polyacetylene and the Peierls mechanism, *Phys. Rev. Lett.* 62, 2016 (1989).

[7] A. Axmann, W. Biem, P. Borsch, F. Hossfeld, and H. Stiller, Lattice dynamics of solid hydrofluoric acid, *Discuss. Faraday Soc.* 48, 69 (1969).

[8] P. Y. Ayala and G. E. Scuseria, Linear scaling second-order Møller–Plesset theory in the atomic orbital basis for large molecular systems, *J. Chem. Phys.* 110, 3660 (1999).

[9] P. Y. Ayala, K. N. Kudin, and G. E. Scuseria, Atomic orbital Laplace-transformed second-order Møller–Plesset theory for periodic systems, *J. Chem. Phys.* 115, 9698 (2001).

[10] R. J. Bartlett, Many-body perturbation theory and coupled cluster theory for electron correlation in molecules, *Ann. Rev. Phys. Chem.* 32, 359 (1981).

[11] R. J. Bartlett and M. Musiał, Coupled-cluster theory in quantum chemistry, *Rev. Mod. Phys.* 79, 291 (2007).

[12] G. J. O. Beran, Approximating quantum many-body intermolecular interactions in molecular clusters using classical polarizable force fields, *J. Chem. Phys.* 130, 164115 (2009).

[13] C. V. Berney and J. W. White, Selective deuteration in neutron-scattering spectroscopy: Formic acid and deuterated derivatives, *J. Am. Chem. Soc.* 99, 6878 (1977).

[14] F. Bloch, Uber die quantenmechanik der elektronen in kristallgittern, *Z. Phys. A* 52, 555 (1928).

[15] S. M. Blumenfeld and H. Fast, Low frequency Raman spectra of solid and liquid formic acid, *Spectrochim. Acta* 24A, 1449 (1968).

[16] H. Boutin, G. J. Safford, and V. Brajovic, Study of low-frequency molecular motions in HF, KHF_2, KH_2F_3, and NaH_2F_3, *J. Chem. Phys.* 39, 3135 (1963).

[17] S. Carter, S. J. Culik, and J. M. Bowman, Vibrational self-consistent field method for many-mode systems: A new approach and application to the vibrations of CO adsorbed on Cu(100), *J. Chem. Phys.* 107, 10458 (1997).

[18] B. Champagne, *Ab initio* polymer quantum theory, in V. Galiatsatos, editor, *Molecular simulation methods for predicting polymer properties*, page 1 (Wiley, New York, 2005).

[19] B. Champagne and J.-M. André, Determination of *ab initio* polarizabilities of polymers: Application to polyethylene and polysilane, *Int. J. Quantum Chem.* 42, 1009 (1992).

[20] M. A. Collins and V. A. Deev, Accuracy and efficiency of electronic energies from systematic molecular fragmentation, *J. Chem. Phys.* 125, 104104 (2006).

[21] E. E. Dahlke and D. G. Truhlar, Electrostatically embedded many-body correlation energy, with applications to the calculation of accurate second-order Møller–Plesset perturbation theory energies for large water clusters, *J. Chem. Theo. Comp.* 3, 1342 (2007).

[22] E. E. Dahlke and D. G. Truhlar, Electrostatically embedded many-body expansion for large systems, with applications to water clusters, *J. Chem. Theo. Comp.* 3, 46 (2007).

[23] V. Deev and M. A. Collins, Approximate *ab initio* energies by systematic molecular fragmentation, *J. Chem. Phys.* 122, 154102 (2005).

[24] G. Del Re, J. Ladik, and G. Biczó, Self-consistent-field tight-binding treatment of polymers. I. Infinite three-dimensional case, *Phys. Rev.* 155, 997 (1967).

[25] J. Delhalle, L. Piela, J. L. Brédas, and J.-M. André, Multipole expansion in tight-binding Hartree–Fock calculations for infinite model polymers, *Phys. Rev. B* 22, 6254 (1980).

[26] J. des Cloizeaux, Energy bands and projection operators in a crystal: Analytic and asymptotic properties, *Phys. Rev.* 135, A685 (1964).

[27] B. I. Dunlap, J. W. D. Connolly, and J. R. Sabin, Some approximations in applications of $X\alpha$ theory, *J. Chem. Phys.* 71, 3396 (1979).

[28] W. Förner, Formulation of the coupled cluster theory with localized orbitals in correlation calculations on polymers, *Int. J. Quantum Chem.* 43, 221 (1992).

[29] W. Förner, R. Knab, J. Čížek, and J. Ladik, Numerical application of the coupled cluster theory with localized orbitals to polymers. IV. Band structure corrections in model systems and polyacetylene, *J. Chem. Phys.* 106, 10248 (1997).

[30] P. Fulde, Wavefunction methods in electronic-structure theory of solids, *Adv. Phys.* 51, 909 (2002).

[31] M. Gadermann, D. Vollmar, and R. Signorell, Infrared spectroscopy of acetic acid and formic acid aerosols: Pure and compound acid/ice particles, *Phys. Chem. Chem. Phys.* 9, 4535 (2007).

[32] S. R. Gadre, R. N. Shirsat, and A. C. Limaye, Molecular tailoring approach for simulation of electrostatic properties, *J. Phys. Chem.* 98, 9165 (1994).

[33] M. S. Gordon, M. A. Freitag, P. Bandyopadhyay, J. H. Jensen, V. Kairys, and W. J. Stevens, The effective fragment potential method: A QM-based MM approach to modeling environmental effects in chemistry, *J. Phys. Chem. A* 105, 293 (2001).

[34] J. Grip and E. J. Samuelsen, Raman spectroscopic search for phase transition in solid formic acid, and lattice dynamics, *Physica Scripta* 24, 52 (1981).

[35] S. Hirata, Fast electron-correlation methods for molecular crystals: An application to the α, β_1, and β_2 polymorphs of solid formic acid, *J. Chem. Phys.* 129, 204104 (2008).

[36] S. Hirata, Quantum chemistry of macromolecules and solids, *Phys. Chem. Chem. Phys.* 11, 8397 (2009).

[37] S. Hirata and R. J. Bartlett, Many-body Green's-function calculations on the electronic excited states of extended systems, *J. Chem. Phys.* 112, 7339 (2000).

[38] S. Hirata, P.-D. Fan, T. Shiozaki, and Y. Shigeta, Single-reference methods for excited states in molecules and polymers, in J. Leszczynski and M. Shukla, editors, *Radiation induced molecular phenomena in nucleic acid: A comprehensive theoretical and experimental analysis*, page 15 (Springer, Berlin, 2008).

[39] S. Hirata, I. Grabowski, M. Tobita, and R. J. Bartlett, Highly accurate treatment of electron correlation in polymers: Coupled-cluster and many-body perturbation theories, *Chem. Phys. Lett.* 345, 475 (2001).

[40] S. Hirata and M. Head-Gordon, Time-dependent density functional theory for radicals: An improved description of excited states with substantial double excitation character, *Chem. Phys. Lett.* 302, 375 (1999).

[41] S. Hirata, M. Head-Gordon, and R. J. Bartlett, Configuration interaction singles, time-dependent Hartree–Fock, and time-dependent density functional theory for the electronic excited states of extended systems, *J. Chem. Phys.* 111, 10774 (1999).

[42] S. Hirata and S. Iwata, Density functional crystal orbital study on the normal vibrations of polyacetylene and polymethineimine, *J. Chem. Phys.* 107, 10075 (1997).

[43] S. Hirata and S. Iwata, Analytical energy gradients in second-order Møller–Plesset perturbation theory for extended systems, *J. Chem. Phys.* 109, 4147 (1998).

[44] S. Hirata and S. Iwata, Analytical second derivatives in *ab initio* Hartree–Fock crystal orbital theory of polymers, *J. Mol. Struct. (Theochem)* 451, 121 (1998).

[45] S. Hirata and S. Iwata, Density functional crystal orbital study on the normal vibrations and phonon dispersion curves of all-*trans* polyethylene, *J. Chem. Phys.* 108, 7901 (1998).

[46] S. Hirata and S. Iwata, *Ab initio* Hartree–Fock and density functional studies on the structures and vibrations of an infinite hydrogen fluoride polymer, *J. Phys. Chem. A* 102, 8426 (1998).

[47] S. Hirata, R. Podeszwa, M. Tobita, and R. J. Bartlett, Coupled-cluster singles and doubles for extended systems, *J. Chem. Phys.* 120, 2581 (2004).

[48] S. Hirata and T. Shimazaki, Fast second-order many-body perturbation method for extended systems, *Phys. Rev. B* 80, 085118 (2009).

[49] S. Hirata, H. Torii, and M. Tasumi, Density-functional crystal orbital study on the structures and energetics of polyacetylene isomers, *Phys. Rev. B* 57, 11994 (1998).

[50] S. Hirata, M. Valiev, M. Dupuis, S. S. Xantheas, S. Sugiki, and H. Sekino, Fast electron correlation methods for molecular clusters in the ground and excited states, *Mol. Phys.* 103, 2255 (2005).

[51] S. Hirata and K. Yagi, Predictive electronic and vibrational many-body methods for molecules and macromolecules, *Chem. Phys. Lett.* 464, 123 (2008).

[52] F. Holtzberg, B. Post, and I. Fankuchen, The crystal structure of formic acid, *Acta Cryst.* 6, 127 (1953).

[53] A. F. Izmaylov and G. E. Scuseria, Resolution of the identity atomic orbital Laplace-transformed second-order Møller–Plesset theory for nonconducting periodic systems, *Phys. Chem. Chem. Phys.* 10, 3421 (2008).

[54] A. F. Izmaylov and G. E. Scuseria, Why are time-dependent density functional theory excitations in solids equal to band structure energy gaps for semilocal functionals, and how does nonlocal Hartree–Fock-type exchange introduce excitonic effects?, *J. Chem. Phys.* 129, 034101 (2008).

[55] D. Jacquemin, J.-M. André, and B. Champagne, Analytic *ab initio* determination of the elastic modulus in stereoregular polymers: Analytical integral derivatives, long-range effects, implementation, and examples, *J. Chem. Phys.* 118, 373 (2003).

[56] D. Jacquemin and B. Champagne, Long-range effects in optimizing the geometry of stereoregular polymers. IV. Explicit determination of the helical angle, *Int. J. Quantum Chem.* 85, 539 (2001).

[57] K. Kamiya, T. Miyamae, M. Oku, K. Seki, H. Inokuchi, C. Tanaka, and J. Tanaka, Ultraviolet photoemission spectra of perchlorate-doped *cis*- and *trans*-polyacetylene, *J. Phys. Chem.* 100, 16213 (1996).

[58] M. Kamiya, S. Hirata, and M. Valiev, Fast electron correlation methods for molecular clusters without basis set superposition errors, *J. Chem. Phys.* 128, 074103 (2008).

[59] A. Karpfen, *Ab initio* studies on polymers. I. Linear infinite polyyne, *J. Phys. C* 12, 3227 (1979).

[60] A. Karpfen, *Ab initio* studies on polymers. V. All-*trans*-polyethylene, *J. Chem. Phys.* 75, 238 (1981).

[61] M. Keçeli and S. Hirata, Fast coupled-cluster singles and doubles for extended systems, unpublished.

[62] M. Keçeli, K. Yagi, and S. Hirata, First-principles methods for anharmonic lattice vibrations: Applications to polyethylene and polyacetylene in the approximation, unpublished.

[63] M. Kertész, Electronic structure of polymers, *Adv. Quantum Chem.* 15, 161 (1982).

[64] B. Kirtman, B. Champagne, F. L. Gu, and D. M. Bishop, Coupled-perturbed Hartree–Fock treatment of infinite periodic systems: Application to static polarizabilities and hyperpolarizabilities of polydiacetylene, polybutatriene, and interacting pairs of polyacetylene chains, *Int. J. Quantum Chem.* 90, 709 (2002).

[65] B. Kirtman, F. L. Gu, and D. M. Bishop, Extension of the Genkin and Mednis treatment for dynamic polarizabilities and hyperpolarizabilities of infinite periodic systems. I. Coupled perturbed Hartree–Fock theory, *J. Chem. Phys.* 113, 1294 (2000).

[66] K. Kitaura, E. Ikeo, T. Asada, T. Nakano, and M. Uebayasi, Fragment molecular orbital method: An approximate computational method for large molecules, *Chem. Phys. Lett.* 313, 701 (1999).

[67] K. Kitaura, T. Sawai, T. Asada, T. Nakano, and M. Uebayasi, Pair interaction molecular orbital method: An approximate computational method for molecular interactions, *Chem. Phys. Lett.* 312, 319 (1999).

[68] J. S. Kittelberger and D. F. Hornig, Vibrational spectrum of crystalline HF and DF, *J. Chem. Phys.* 46, 3099 (1967).

[69] W. Koch and M. C. Holthausen, *A chemist's guide to density functional theory*, 2nd edition (Wiley-VCH, Weinheim, 2001).

[70] K. N. Kudin and G. E. Scuseria, Analytic stress tensor with the periodic fast multipole method, *Phys. Rev. B* 61, 5141 (2000).

[71] A. B. Kunz, Electronic polarons in nonmetals, *Phys. Rev. B* 6, 606 (1972).

[72] J. J. Ladik, *Quantum theory of polymers as solids* (Plenum, New York, 1988).

[73] S. H. Li, J. Shen, W. Li, and Y. S. Jiang, An efficient implementation of the "cluster-in-molecule" approach for local electron correlation calculations, *J. Chem. Phys.* 125, 074109 (2006).

[74] C. M. Liegener, Third-order many-body perturbation theory in the Møller–Plesset partitioning applied to an infinite alternating hydrogen chain, *J. Phys. C* 18, 6011 (1985).

[75] C. M. Liegener, *Ab initio* calculations of correlation effects in *trans*-polyacetylene, *J. Chem. Phys.* 88, 6999 (1988).

[76] F. R. Manby, D. Alfè, and M. J. Gillan, Extension of molecular electronic structure methods to the solid state: Computation of the cohesive energy of lithium hydride, *Phys. Chem. Chem. Phys.* 8, 5178 (2006).

[77] N. H. March, W. H. Young, and S. Sampanthar, *The many-body problem in quantum mechanics* (Cambridge University Press, Cambridge, 1967).

[78] Y. Mikawa, J. W. Brasch, and R. J. Jakobsen, Infrared spectra and normal coordinate calculation of crystalline formic acid, *J. Mol. Spectrosc.* 24, 314 (1967).

[79] Y. Mikawa, R. J. Jakobsen, and J. W. Brasch, Infrared evidence of polymorphism in formic acid crystals, *J. Chem. Phys.* 45, 4750 (1966).

[80] R. C. Millikan and K. S. Pitzer, The infrared spectra of dimeric and crystalline formic acid, *J. Am. Chem. Soc.* 80, 3515 (1958).

[81] J. W. Mintmire and J. R. Sabin, Local density functional methods in two-dimensionally periodic systems. I. The atomic hydrogen monolayer, *Int. J. Quantum Chem. Symp.* 18, 707 (1980).

[82] J. W. Mintmire, J. R. Sabin, and S. B. Trickey, Local-density-functional methods in two-dimensionally periodic systems: Hydrogen and beryllium monolayers, *Phys. Rev. B* 26, 1743 (1982).

[83] J. W. Mintmire and C. T. White, Local-density-functional results for the dimerization of *trans*-polyacetylene: Relationship to the band-gap problem, *Phys. Rev. B* 35, 4180 (1987).

[84] H. J. Monkhorst and M. Kertesz, Exact-exchange asymptotics in polymer Hartree–Fock calculations, *Phys. Rev. B* 24, 3015 (1981).

[85] I. Nahringbauer, A reinvestigation of the structure of formic acid (at 98 K), *Acta Cryst. B* 34, 315 (1978).

[86] J. Paloheimo and J. von Boehm, Density-functional study of the dimerization of *trans*-polyacetylene, *Phys. Rev. B* 46, 4304 (1992).

[87] S. F. Parker, Inelastic neutron scattering spectra of polyethylene, *J. Chem. Soc. Faraday Trans.* 92, 1941 (1996).

[88] R. G. Parr and W. Yang, *Density-functional theory of atoms and molecules.* (Oxford University Press, Oxford, 1989).

[89] B. Paulus, The method of increments—A wavefunction-based ab initio correlation method for solids, *Int. J. Mod. Phys. B* 21, 2204 (2007).

[90] L. Piela, J.-M. André, J. G. Fripiat, and J. Delhalle, On the behavior of exchange in restricted Hartree–Fock–Roothaan calculations for periodic polymers, *Chem. Phys. Lett.* 77, 143 (1981).

[91] R. Pino and G. E. Scuseria, Laplace-transformed diagonal Dyson correction to quasiparticle energies in periodic systems, *J. Chem. Phys.* 121, 2553 (2004).

[92] R. Pino and G. E. Scuseria, Importance of chain-chain interactions on the band gap of *trans*-polyacetylene as predicted by second-order perturbation theory, *J. Chem. Phys.* 121, 8113 (2004).

[93] A. Pomogaeva, B. Kirtman, F. L. Gu, and Y. Aoki, Band structure built from oligomer calculations, *J. Chem. Phys.* 128, 074109 (2008).

[94] T. D. Poulsen, K. V. Mikkelsen, J. G. Fripiat, D. Jacquemin, and B. Champagne, MP2 correlation effects upon the electronic and vibrational properties of polyyne, *J. Chem. Phys.* 114, 5917 (2001).

[95] P. Pulay, Analytic derivative methods in quantum chemistry, in K. P. Lawley, editor, *Advances in chemical physics, Vol. 69: Ab initio methods in quantum chemistry, Part II,* page 241 (Wiley, New York, 1987).

[96] P. Reinhardt, Dressed coupled-electron-pair-approximation methods for periodic systems, *Theor. Chem. Acc.* 104, 426 (2000).

[97] A. L. Ringer and C. D. Sherrill, First principles computation of lattice energies of organic solids: The benzene crystal, *Chem. Eur. J.* 14, 2542 (2008).

[98] J. Sauer, Molecular models in ab initio studies of solids and surfaces: From ionic crystals and semiconductors to catalysts, *Chem. Rev.* 89, 199 (1989).

[99] I. Shavitt, The history and evolution of configuration interaction, *Mol. Phys.* 94, 3 (1998).

[100] T. Shimazaki and S. Hirata, On the Brillouin-zone integrations in second-order many-body perturbation calculations for extended systems of one-dimensional periodicity, *Int. J. Quantum Chem.* 109, 2953 (2009).

[101] T. Shiozaki, M. Kamiya, S. Hirata, and E. F. Valeev, Equation of explicitly-correlated coupled-cluster methods, *Phys. Chem. Chem. Phys.* 10, 3358 (2008).

[102] T. Shiozaki, M. Kamiya, S. Hirata, and E. F. Valeev, Explicitly correlated coupled-cluster singles and doubles method based on complete diagrammatic equations, *J. Chem. Phys.* 129, 071101 (2008).

[103] T. Shiozaki, M. Kamiya, S. Hirata, and E. F. Valeev, Higher-order explicitly correlated coupled-cluster methods, *J. Chem. Phys.* 130, 054101 (2009).

[104] R. G. Snyder and J. H. Schachtschneider, Vibrational analysis of the *n*-paraffins. I. Assignments of infrared bands in the spectra of C_3H_8 through *n*-$C_{19}H_{40}$, *Spectrochim. Acta* 19, 85 (1963).

[105] O. Sode, M. Keçeli, S. Hirata, and K. Yagi, Coupled-cluster and many-body perturbation study of energies, structures, and phonon dispersions of solid hydrogen fluoride, *Int. J. Quantum Chem.* 109, 1928 (2009).

[106] H. Stoll, B. Paulus, and P. Fulde, On the accuracy of correlation-energy expansions in terms of local increments, *J. Chem. Phys.* 123, 144108 (2005).

[107] S. Suhai, Calculation of the mechanical and optical properties of polyethylene including electron correlation effects, in J. Ladik, J.-M. André, and M. Seel, editors, *Quantum chemistry of polymers: Solid state aspects* (Reidel, Braunlage, 1983).

[108] S. Suhai, Bond alternation in infinite polyene: Peierls distortion reduced by electron correlation, *Chem. Phys. Lett.* 96, 619 (1983).

[109] S. Suhai, Quasiparticle energy-band structures in semiconducting polymers: Correlation effects on the band gap in polyacetylene, *Phys. Rev. B* 27, 3506 (1983).

[110] S. Suhai, Electron correlation effects on the mechanical and optical properties of polymers, *Int. J. Quantum Chem. Symp.* 18, 161 (1984).

[111] S. Suhai, Green's-function study of optical properties of polymers: Charge-transfer exciton spectra of polydiacetylenes, *Phys. Rev. B* 29, 4570 (1984).

[112] S. Suhai, On the excitonic nature of the first UV absorption peak in polyene, *Int. J. Quantum Chem.* 29, 469 (1986).

[113] S. Suhai, Theory of exciton–photon interaction in polymers: Polariton spectra of polydiacetylenes, *J. Chem. Phys.* 85, 611 (1986).

[114] S. Suhai, Electron correlation in extended systems: Fourth-order many-body perturbation theory and density-functional methods applied to an infinite chain of hydrogen atoms, *Phys. Rev. B* 50, 14791 (1994).

[115] S. Suhai, Cooperative effects in hydrogen bonding: Fourth-order many-body perturbation theory studies of water oligomers and of an infinite water chain as a model for ice, *J. Chem. Phys.* 101, 9766 (1994).

[116] S. Suhai, Electron correlation and dimerization in *trans*-polyacetylene: Many-body perturbation theory versus density-functional methods, *Phys. Rev. B* 51, 16553 (1995).

[117] J.-Q. Sun and R. J. Bartlett, Second-order many-body perturbation-theory calculations in extended systems, *J. Chem. Phys.* 104, 8553 (1996).

[118] J.-Q. Sun and R. J. Bartlett, Correlated prediction of the photoelectron spectrum of polyethylene: Explanation of XPS and UPS measurements, *Phys. Rev. Lett.* 77, 3669 (1996).

[119] J.-Q. Sun and R. J. Bartlett, Many-body perturbation theory for quasiparticle energies, *J. Chem. Phys.* 107, 5058 (1997).

[120] J.-Q. Sun and R. J. Bartlett, Convergence of many-body perturbation methods with lattice summations in extended systems, *J. Chem. Phys.* 106, 5554 (1997).

[121] J.-Q. Sun and R. J. Bartlett, Analytical evaluation of energy derivatives in extended systems. I. Formalism, *J. Chem. Phys.* 109, 4209 (1998).

[122] J.-Q. Sun and R. J. Bartlett, Convergence behavior of many-body perturbation theory with lattice summations in polymers, *Phys. Rev. Lett.* 80, 349 (1998).

[123] J.-Q. Sun and R. J. Bartlett, Modern correlation theories for extended, periodic systems, *Topics Current Chem.* 203, 129 (1999).

[124] M. Tasumi and T. Shimanouchi, Crystal vibrations and intermolecular forces of polymethylene crystals, *J. Chem. Phys.* 43, 1245 (1965).

[125] M. Tasumi, T. Shimanouchi, and T. Miyazawa, Normal vibrations and force constants of polymethylene chain, *J. Mol. Spectrosc.* 9, 261 (1962).

[126] M. Tasumi, T. Shimanouchi, and T. Miyazawa, A refined treatment of normal vibrations of polymethylene chain, *J. Mol. Spectrosc.* 11, 422 (1963).

[127] H. Teramae, T. Yamabe, and A. Imamura, *Ab initio* studies on the geometrical and vibrational structures of polymers, *J. Chem. Phys.* 81, 3564 (1984).

[128] H. Teramae, T. Yamabe, C. Satoko, and A. Imamura, Energy gradient in the ab initio Hartree–Fock crystal-orbital formalism of one-dimensional infinite polymers, *Chem. Phys. Lett.* 101, 149 (1983).

[129] M. Tobita, S. Hirata, and R. J. Bartlett, The analytical energy gradient scheme in the Gaussian based Hartree–Fock and density functional theory for two-dimensional systems using fast multipole method, *J. Chem. Phys.* 118, 5776 (2003).

[130] G. S. Tschumper, Multicentered integrated QM:QM methods for weakly bound clusters: An efficient and accurate 2-body:many-body treatment of hydrogen bonding and van der Waals interactions, *Chem. Phys. Lett.* 427, 185 (2006).

[131] R. Tubino and G. Zerbi, Phonon curves and frequency spectrum for hydrogen-bonded systems: Solid HF and DF, *J. Chem. Phys.* 51, 4509 (1969).

[132] R. Tubino and G. Zerbi, Phonon curves and frequency distribution for hydrogen-bonded systems: β modification of solid HCOOH, HCOOD, and DCOOH, *J. Chem. Phys.* 53, 1428 (1970).

[133] P. Vogl and D. K. Campbell, Three-dimensional structure and intrinsic defects in *trans*-polyacetylene, *Phys. Rev. Lett.* 62, 2012 (1989).

[134] P. Vogl and D. K. Campbell, First-principles calculations of the three-dimensional structure and intrinsic defects in *trans*-polyacetylene, *Phys. Rev. B* 41, 12797 (1990).

[135] M. Vračko, B. Champagne, D. H. Mosley, and J. M. André, Study of excited states of polyethylene in the Hartree–Fock, Tamm–Dancoff, and random-phase approximations, *J. Chem. Phys.* 102, 6831 (1995).

[136] M. G. Vračko and M. Zaider, A calculation of exciton energies in periodic systems with helical symmetry: Application to a hydrogen fluoride chain, *Int. J. Quantum Chem.* 43, 321 (1992).

[137] M. G. Vračko and M. Zaider, A study of excited-states in *trans*-polyacetylene in the Hartree–Fock, Tamm–Dancoff, and random-phase approximation, *Int. J. Quantum Chem.* 47, 119 (1993).

[138] K. Yagi, S. Hirata, and K. Hirao, Efficient configuration selection scheme for vibrational second-order perturbation theory, *J. Chem. Phys.* 127, 034111 (2007).

[139] Y.-J. Ye, W. Förner, and J. Ladik, Numerical application of the coupled-cluster theory with localized orbitals to polymers. I. Total correlation energy per unit cell, *Chem. Phys.* 178, 1 (1993).

[140] D. Yoshimura, H. Ishii, Y. Ouchi, E. Ito, T. Miyamae, S. Hasegawa, K. K. Okudaira, N. Ueno, and K. Seki, Angle-resolved ultraviolet photoelectron spectroscopy and theoretical simulation of a well-ordered ultrathin film of tetratetracontane (n-$C_{44}H_{90}$) on Cu(100): Molecular orientation and intramolecular energy-band dispersion, *Phys. Rev. B* 60, 9046 (1999).

[141] H. R. Zelsmann, F. Bellon, Y. Marechal, and B. Bullemer, Polymorphism in crystalline formic acid, *Chem. Phys. Lett.* 6, 513 (1970).

[142] D. W. Zhang and J. Z. H. Zhang, Molecular fractionation with conjugate caps for full quantum mechanical calculation of protein-molecule interaction energy, *J. Chem. Phys.* 119, 3599 (2003).

chapter seven

Ab initio Monte Carlo simulations of liquid water

Darragh P. O'Neill, Neil L. Allan, and Frederick R. Manby

Contents

7.1 Introduction

Liquids pose a difficult challenge for theorists. On the one hand, they require an accurate description of the statistical mechanics of the system for bulk behavior, and on the other hand, they require modeling of the quantum-mechanical interactions between individual particles in the system. Approximations of both the statistical mechanics and the quantum mechanics are needed. The simplest approach would be to take clusters of molecules that one hopes will be representative of the bulk. Considering only hundreds or even thousands of particles may, however, still be far from capturing bulk conditions and the effect of any boundary may be large. Performing an ab initio simulation on large clusters is unfeasible and is more likely limited to much smaller clusters of around one hundred atoms. A better approach is to use periodic boundary conditions [1] in which a box of particles is replicated across all space. Interactions between boxes must be considered, but only the particles in a single box are treated explicitly. This gives a more realistic set of boundary conditions at more reasonable computational expense, because the box can be relatively small. Although the use of periodic boundary conditions is not without its problems because an artificial periodicity is imposed, we will not focus on this approximation, as we are primarily concerned with improving the treatment of the molecular interactions.

Traditionally, most condensed-phase calculations have been performed by approximating the interactions between particles with empirical potentials. These have the advantage of being very computationally efficient as, depending on the sophistication of the potential, the evaluation of each interaction requires only a few arithmetic operations. Potentials vary in their sophistication from the very simple Lennard–Jones potentials [2], which crudely describe a short-range repulsive potential with a longer-range attractive potential, to much more sophisticated potentials such as those of Bukowski and co-workers [3,4] in which interaction potentials are derived from symmetry-adapted perturbation theory [5] and fitted to highly accurate ab initio results. There are myriad potentials (Guillot lists more than 40 for water alone [6]) whose functional forms have been tailored, with parameters fitted to specific systems or to reproduce certain properties, and although tremendously popular and often reasonably successful, they suffer from the drawback that the potentials must often be refitted when a different system is considered or when the conditions of interest are far from those sampled in the parameterization [7]. If a more accurate model is desired, it is not always clear how potentials can be improved and the development of a new potential can be a labor-intensive task and often relies on empiricism as opposed to rigorous justification.

Another widely used approach is density-functional theory (DFT) [8]. DFT is, in principle, an ab initio method and, due to its computational efficiency, can be applied to large systems and also to periodic systems. It has been applied to both solids [9] and liquids [10] and in many situations works well. However, although DFT is an ab initio theory, it requires, in practice, a choice to be made of a functional form and parameterization of the exchange-correlation functional. There are a huge number of functionals and it is often unclear which functional to use for a particular system; a functional that is successful in one situation may not work well in others. It is also not clear how to improve density functionals systematically.

Wave-function-based electronic-structure theory has a well-established hierarchy of model chemistries (a given method using a given basis set) [11]. As these methodologies are based on a series of well-defined approximations, it is always possible to achieve arbitrary accuracy by systematically improving the method and basis set; but this is, of course, accompanied by additional computational expense. In comparison to potential-based methods or DFT, wavefunction-based methods are computationally much more demanding and have been primarily applied to small- to medium-sized systems in vacuo. A notable exception to this is the work of Kresse and co-workers who have developed a periodic implementation of MP2 theory [12] and are developing a similar coupled-cluster methodology. Another exception is the periodic local MP2 implementation of Pisani et al. [13] and Laplace-transformed MP2 implementation of Ayala et al. [14]. Although noteworthy developments, these periodic ab initio methods are very complicated and computationally expensive. Our goal is to develop a simple approach to allow the use of the hierarchy of quantum-chemical methods in condensed phases and provide a systematically improvable, nonempirical, and computationally accessible approach to these challenging problems.

In particular this work will focus on the simulation of liquid water. Simulations of water have spanned four decades [6], using a host of different methods to varying levels of success, but none has used a systematically improvable ab initio methodology. The most frequently used methods are potential based, and potentials such as TIP3P, TIP4P [15], and SPC/E [16] are particularly popular. These models contain numerous parameters, such as point charges on the molecules and Lennard–Jones parameters, which are fitted to reproduce experimental data. Although these models reproduce some properties well, outside of their parameterization sets they may perform badly. TIP4P, for example, reproduces the density under ambient conditions, since it is fitted to do so, but incorrectly predicts the density maximum to lie within the supercooled region of the phase diagram [17]. As mentioned above, potentials can vary greatly in their degree of sophistication, and many of those for water are based on nonflexible and nonpolarizable water molecules and include only two-body interactions. It is known, however, that many-body effects, for example, are physically important phenomena in liquid water, and in these simpler potentials, such effects are accounted for by effective potentials that are suitable only for a specific region of the phase diagram.

There are other potentials including monomer flexibility, polarization, and 3-body interactions, and particularly noteworthy is that of Bukowski and co-workers [3,4], which is derived from ab initio data and includes both many-body effects and polarization effects, although without monomer flexibility. This potential reproduces various experimental data ranging from the dimer to the condensed phase. Though this potential has been rigorously derived from ab initio data, to improve it or to apply it to other systems, it must be reformulated and reparameterized.

DFT has also been used to investigate liquid water since the first pioneering simulations using Car–Parrinello molecular dynamics (CPMD) in 1993 [18]. Since then numerous DFT studies (e.g., [19–27]) have provided powerful insights into the properties of this fundamental system. DFT simulations have remained the theoretically most rigorous approach to be applied to liquid water, and although extremely successful, DFT has certain shortcomings and does not, for example, correctly describe the structural properties of water under ambient conditions, predicting an overstructured radial-distribution function and a diffusion constant that is an order of magnitude too small [22]. The reasons for this are various and include the intrinsic error of the functional being used, the small sizes of the cells used, a poor description of the intramolecular distortions, and the fictitious electronic mass used in Car–Parrinello molecular dynamics [22, 28].

It is clear that potential-based methods and DFT do not always perform satisfactorily, and this coupled with a lack of systematic improvability suggests that it would be desirable to use wavefunction-based methods to describe condensed phases. Many attempts to do this have used the many-body expansion (MBE), dividing the total energy into sums of dimer, trimer, etc. energies, and the MBE will also be used here. Truhlar and coworkers use the MBE embedded in the electrostatic environment of the complete system and have applied this to clusters of water molecules [29–31] (see also Chapter 5 in this volume). Work by Beran also uses the MBE but in conjunction with an empirical polarizable force field to account for many-body effects, and he has used this to investigate water clusters [32].

The broad philosophy we use to tackle the problem of applying quantum-chemical methods to condensed phases is to simplify the problem by decomposing it into its constituent parts. We then deal with each of these components separately, treating them with the most computationally efficient, but also the most theoretically appropriate way, while always ensuring that we retain systematic improvability. We also employ the MBE to reduce the problem to a set of calculations on dimers, trimers, etc., and we treat higher-order effects through a polarizable model that uses ab initio properties. We also make a spatial partitioning of these smaller clusters, which allows us to treat clusters differently, depending on whether they interact strongly or weakly.

The methodology we describe in the next sections is not specific to water and can be applied to any molecular liquid or solid, and furthermore can be used with any quantum-chemical method or basis set, as well as being straightforward to parallelize. We will present results on small to medium water clusters as well as on liquid water.

7.2 Theory

7.2.1 Many-body expansion

First, we simplify the problem by invoking the many-body expansion (MBE). Instead of considering the system interacting as a whole we consider it as a collection of two-, three-, etc., body interactions, as follows

$$E = \sum_i E_i + \sum_{i<j} \delta E_{ij} + \sum_{i<j<k} \delta E_{ijk} + \dots , \qquad (7.1)$$

where

$$\delta E_{ij} = E_{ij} - E_i - E_j \qquad (7.2)$$

$$\delta E_{ijk} = E_{ijk} - \delta E_{ij} - \delta E_{jk} - \delta E_{ik} - E_i - E_j - E_k. \qquad (7.3)$$

Equation (7.1) is exact, as the energy of the whole interacting system, $E_{ij..n}$ is included. It is hoped that this expansion converges quickly and can be truncated without significant loss of accuracy. Although the majority of potential-based methods explicitly include only pairwise interactions, the inclusion of higher-order effects is essential in many systems such as water.

To investigate this further we calculate the low-order contributions to the many-body expansion for some small water clusters, in a manner similar to that in previous studies [33,34]. The results are shown in Table 7.1. Although the two-body energy dominates, the three-body contributions are significant at approximately 3 mE_h per water molecule and even the four-body contributions cannot be neglected. It can, however, be seen that the expansion is converging and that much higher-order contributions will not be significant.

7.2.2 Spatial partitioning of interactions

These interaction energies can be further decomposed into those involving monomers that are close and those that are farther apart. For example, for

Table 7.1 DF-LMP2/AVDZ Two-, Three-
and Four-Body Contributions per Water
Molecule to the Energies of Clusters of n
Water Molecules. All Energies are in mE_h

n	2-body	3-body	4-body
6	−7.9	−2.5	−0.4
8	−10.6	−2.4	−0.2
10	−10.6	−2.7	−0.5
12	−10.7	−2.8	−0.6
14	−10.9	−2.9	−0.7
16	−11.0	−2.9	−0.7
18	−11.4	−3.0	−0.8
20	−11.7	−3.0	−1.0

the two-body interaction energy we have

$$\sum_{i<j} \delta E_{ij} = \sum_{\substack{i<j \\ i,\, j\,\text{near}}} \delta E_{ij} + \sum_{\substack{i<j \\ i,\, j\,\text{far}}} \delta E_{ij}. \tag{7.4}$$

Similar expressions can be written down for higher-order interactions. A threshold distance must be chosen to categorise clusters as either near or far and the choice of this parameter will be discussed in Section 7.3. Interactions between monomers that are close together will be influenced strongly by quantum-mechanical effects, whereas those that are well separated and where there is no significant overlap of their charge distributions will interact in a more classical way. The largest quantum-mechanical effect is exchange that decays exponentially with distance, so only very close monomers should need to be treated with expensive QM methods, while cheaper methods are appropriate for those farther away. This approximation can always be improved by simply increasing the threshold distance, within which more accurate methods are used.

7.2.3   Quantum-mechanical description of interactions

As mentioned in the previous section, interactions between monomers whose charge distributions overlap must be calculated using techniques that adequately include quantum-mechanical effects. In the methodology described here, any method can be used and the usual trade-off between computational cost and accuracy must be made. In this work we focus on post-Hartree–Fock methods that treat electron correlation [11], but any others including potential-based methods and DFT could be used.

7.2.3.1   Basis-set superposition error

When calculating interaction energies using methods that require a finite basis set, care must be taken as basis-set superposition error (BSSE) can cause problems. This results from the use of a finite basis set and the simplest way to ameliorate it is to use a larger basis. This is unfortunately not always possible due to the increased computational cost, so there are other methods to combat this problem, although some are more appropriate than others within this scheme.

The most frequently used solution to BSSE is counterpoise correction [35], in which the supermolecule basis set is used for all calculations; i.e., for a dimer calculation the monomer energies are calculated in the basis set of the dimer. The counterpoise correction usually works quite well, but it should not be used in conjunction with the MBE. For the MBE to be exact it relies on the exact cancellation of lower-order terms, but if different interactions have been calculated in different basis sets, this cancellation is not exact. The only solution to this would be to do all calculations in the basis set for the complete cluster. Normally, however, this would not be computationally feasible beyond small clusters.

Another possible solution to BSSE is the use of local methods, although we note that this does not address BSSE in the Hartree–Fock portion of the energy. In these methods the orbitals are localized on the monomers and the correlation treatment is constrained to omit double excitations from orbitals on one monomer into the virtual space of another monomer [36]. This corrects the larger part of the BSSE, and is the method we use in this work.

7.2.4   Classical description of interactions

When the monomers are far enough apart, there is no significant overlap of their charge distributions and the interaction can be described classically. The first step is to decompose this energy further

$$E = E_{es} + E_{ind} + E_{disp} + E_{ex,} \qquad (7.5)$$

where the four terms on the right-hand side are the electrostatic, induction, dispersion, and exchange energies, respectively. Each of these can then be expressed in terms of properties specific to the separate monomers and the distance between the monomers. This will be discussed briefly here, but for a complete discussion of these interactions and how they arise, we refer the reader to [37].

7.2.4.1   Monomer properties

The classical interactions are evaluated in terms of monomer properties, namely the multipole moments and polarizability tensors, which provide a compact description of the charge density and of the response of this density to an external field, respectively. The multipole moments are defined as

$$Q_{l\kappa} = \int \rho(\mathbf{r}) R_{l\kappa}(\mathbf{r}) d\mathbf{r}, \tag{7.6}$$

where $\rho(\mathbf{r})$ is the charge density and $R_{l\kappa}(\mathbf{r})$ are the regular real spherical harmonics [37]; the subscript l is angular momentum and κ is one of the corresponding $(2l+1)$ components. The polarizability is then given as the linear response of the multipole moments to an external field

$$\alpha_{l\kappa,l'\kappa'} = -\left(\frac{dQ_{l\kappa}}{dV_{l'\kappa'}}\right)_{V\to 0}, \tag{7.7}$$

where $V_{l\kappa}$ is a component of the external field. Both the multipole moments and the polarizability tensor can be calculated in almost all electronic structure program packages. We also note that both quantities may be calculated in terms of Cartesian components instead of in terms of spherical harmonics, but this leads to a less efficient and also a less compact formulation when calculating the various interactions.

Equation (7.6) for the multipole moments provides a very compact description of the charge density of a molecule, where we have now reduced a complicated density to a simple expansion about a single center. Although very compact, this may not be very good, especially in large molecules or when evaluating the potential close to the molecule; here, multipole moments of very large angular momentum l may be required to provide an adequate description. To circumvent this problem it is possible to perform a so-called *distributed multipole analysis* [38–40] in which the density is not expanded in moments around a single center, but rather around multiple centers, usually, but not necessarily, chosen to lie at the atomic positions. Similar schemes have been devised for distributed polarizabilities [41–43]. In the subsequent sections all formulae for interactions will be presented in terms of a single multipole and polarizable site per monomer for the sake of simplicity, but the use of distributed properties is straightforward the relevant formulae are given in [37], and in our implementation we use distributed properties.

7.2.4.2   *Electrostatic energy*

The electrostatic energy is the classical Coulomb interaction between two charge distributions. It decays as an inverse power of the distance between monomers, R, and is exactly two-body additive, i.e., it can always be reduced to a sum of pairwise interactions. The total electrostatic energy for a system is given by

$$E_{es} = \frac{1}{2} \sum_{AB} \sum_{l_a \kappa_a} \sum_{l_b \kappa_b} Q^A_{l_a \kappa_a} T^{AB}_{l_a \kappa_a, l_b \kappa_b} Q^B_{l_b \kappa_b}, \tag{7.8}$$

where $Q^M_{l \kappa}$ is the multipole moment on a molecule M of angular momentum l and component κ, and $T^{AB}_{l_a \kappa_a, l_b \kappa_b}$ is a component of the interaction tensor, which scales as $1/R^{l_a + l_b + 1}$. Evaluation of the interaction tensor will be discussed in Section 7.2.4.6.

7.2.4.3 Induction energy

The induction energy is the energy change caused by the distortion of the charge density by the field due to the other molecules and is given by

$$E_{ind} = \frac{1}{2} \sum_{AB} \sum_{l_a \kappa_a} \sum_{l_b \kappa_b} \Delta Q^A_{l_a \kappa_a} T^{AB}_{l_a \kappa_a, l_b \kappa_b} Q^B_{l_b \kappa_b}, \tag{7.9}$$

$$V^A_{l\kappa} = \sum_{B} \sum_{l_b \kappa_b} T^{AB}_{l\kappa, l_b \kappa_b} Q^B_{l_b \kappa_b}, \tag{7.10}$$

$$\Delta Q^A_{l_a \kappa_a} = - \sum_{l\kappa} \alpha^A_{l_a \kappa_a, l\kappa} V^A_{l\kappa}, \tag{7.11}$$

where $\Delta Q^A_{l_a \kappa_a}$ are the multipoles induced on A, and $\alpha^A_{l_a \kappa_a, l_b \kappa_b}$ is an element of the polarizability tensor as defined in Equation (7.7). The induction interaction decays as an inverse power of R, but is not two-body additive. This is because the induced multipoles are caused by the field due to all other molecules; these in turn change the field and thus induce a different set of multipoles, and so on. Thus the induction energy must be evaluated iteratively and cannot be reduced to a simple sum of pairwise interactions. The importance of iterating this can be seen in a cluster of 50 water molecules where the difference between the noniterated and iterated induction energies is 1.6 mE_h per water molecule. The induction energy is the major component of the many-body energy and will be used in this work to estimate the total many-body energy, where QM calculations are not used.

7.2.4.4 Dispersion energy

The dispersion interaction arises from correlated fluctuations in the electronic densities of the two molecules. It decays as an inverse power of R and is approximately two-body additive. Although this is a quantum-mechanical effect, describing the correlated motion of the electrons, it can be expanded in the following series

$$E_{\text{disp}} = -\sum_{AB} \sum_{n=6}^{\infty} \sum_{l_a \kappa_a, l_b \kappa_b, j} \frac{C_n(l_a, l_b, j, \kappa_a, \kappa_b)}{R^n} \bar{S}_{l_a, l_b, j}^{\kappa_a \kappa_b}(\Omega_A, \Omega_B, \Omega_{AB}), \quad (7.12)$$

where $C_n(l_a, l_b, j, \kappa_a, \kappa_b)$ are the dispersion coefficients and $\bar{S}_{l_a, l_b, j}^{\kappa_a \kappa_b}(\Omega_A, \Omega_B, \Omega_{AB})$ are Stone's orientational functions [37]. These are related to the interaction tensor and will be discussed in Section 7.2.4.6. The dispersion coefficients can be calculated through the frequency-dependent polarizabilities of the two species involved [37,44–46]. A simpler isotropic form of Equation (7.12) can also be used and is given by

$$E_{\text{disp}} = -\sum_{n=6}^{10} \frac{C_n}{R^n}, \quad (7.13)$$

where the isotropic dispersion coefficients are $C_{2n} = C_{2n}(n, n, 0, 0, 0)$. This form of the dispersion energy usually suffices when the monomers are not close together. We note that, as for the multipole moments and polarizabilities, a distributed dispersion model can also be formulated [41,47,48].

7.2.4.5 Exchange energy

The exchange interaction is a purely quantum-mechanical effect and arises from the exchange of electrons between the two monomers. It decays exponentially with R and is not two-body additive. As it decays so rapidly, it is usually negligible for molecules that are well separated, but can be estimated using the overlap of the charge densities [49]

$$E_{\text{ex}} = K S_{AB}, \quad (7.14)$$

where S_{AB} is the overlap between molecules A and B and K is a parameter. S_{AB} can be evaluated using density fitting [50].

7.2.4.6 Calculation of the interaction tensor

In the preceeding sections the interaction tensor **T** has been mentioned several times. The interaction tensor describes the interaction between two sets of multipoles and depends on the relative orientations of the two molecules with respect to one another, but also with respect to some reference frame in which the multipoles have been calculated. A more detailed derivation of the interaction tensor can be found in [37]. To calculate the interaction tensor and the various interactions we follow Hättig [51] and write the interaction tensor as

$$T_{l_a \kappa_a, l_b \kappa_b}^{AB} = R_{AB}^{-l_a - l_b - 1} \sum_{\kappa} D_{\kappa \kappa_a}^{(l_a)} \left(\Omega_{AB}^{-1} \Omega_A \right) t_{\kappa}^{(l_a, l_b)} D_{\kappa \kappa_b}^{(l_b)} \left(\Omega_{AB}^{-1} \Omega_B \right), \quad (7.15)$$

where Ω_A and Ω_B define the orientations of the molecules with respect to their reference frames, and Ω_{AB} the relative orientation of the two molecules; $\mathbf{D}$ are Wigner rotation matrices and $t_\kappa^{(l_a,l_b)}$ are a set of coefficients. These rotations align the multipoles of the two molecules along the z-axis and this results in the coefficients $t_\kappa^{(l_a,l_b)}$ taking a very simple form, given by

$$t_\kappa^{(l_a,l_b)} = (-1)^{l_b} \sqrt{\binom{l_a + l_b}{l_a + \kappa}\binom{l_a + l_b}{l_a - \kappa}}. \tag{7.16}$$

In our work the monomers are allowed to distort, and as such there is no advantage to having the properties in some reference frame since the properties change with molecular geometry. They are instead calculated on the fly in the global frame, and thus Ω_A and Ω_B are not required and

$$T_{l_a\kappa_a,l_b\kappa_b}^{AB} = R_{AB}^{-l_a-l_b-1} \sum_\kappa D_{\kappa\kappa_a}^{(l_a)}\left(\Omega_{AB}^{-1}\right) t_\kappa^{(l_a,l_b)} D_{\kappa\kappa_b}^{(l_b)}\left(\Omega_{AB}^{-1}\right). \tag{7.17}$$

Only a single set of Wigner rotation matrices need be calculated for each AB pair. To calculate an electrostatic interaction energy we take the following steps [51]:

$$Q_{l_b\kappa}^{B,r} = \sum_{\kappa_b} D_{\kappa\kappa_b}^{(l_b)}\left(\Omega_{AB}^{-1}\right) Q_{l_b\kappa_b}^{B}, \tag{7.18}$$

$$V_{l_a\kappa}^{A,r} = \sum_{l_b} R_{AB}^{-l_a-l_b-1} t_\kappa^{(l_a,l_b)} Q_{l_b\kappa}^{B,r}, \tag{7.19}$$

$$V_{l_a\kappa_a}^{A} = \sum_\kappa D_{\kappa\kappa_a}^{(l_a)}\left(\Omega_{AB}^{-1}\right) V_{l_a\kappa}^{A,r}, \tag{7.20}$$

$$E_{\text{es}} = \sum_{l_a\kappa_a} Q_{l_a\kappa_a}^{A} V_{l_a\kappa_a}^{A}, \tag{7.21}$$

where the superscript r denotes in the rotated frame. To evaluate the Wigner rotation matrices efficiently we use the recursive method described by Ivanic and Ruedenberg [52,53].

7.2.5   Self-consistent induction calculations

As mentioned in Section 7.2.4.3 the induction energy must be evaluated self-consistently. As in Equation (7.11), the induced multipoles are given by

$$\Delta Q_{l_a\kappa_a}^{A} = -\sum_{l\kappa} \alpha_{l_a\kappa_a,l\kappa}^{A} V_{l\kappa}^{A}. \tag{7.22}$$

These induced multipoles change the field experienced by the other sites, leading to another set of induced multipole moments

$$\Delta V_{l_a \kappa_a}^A = \sum_B \sum_{l_b \kappa_b} T_{l_a \kappa_a, l_b \kappa_b}^{AB} \Delta Q_{l_b \kappa_b}^B, \tag{7.23}$$

$$\Delta \Delta Q_{l_a \kappa_a}^A = - \sum_{l\kappa} \alpha_{l_a \kappa_a, l\kappa}^A \Delta V_{l\kappa}^A, \tag{7.24}$$

and so on until convergence. This can be written as a set of coupled equations

$$\Delta Q_{l_a \kappa_a}^A = \sum_{l\kappa} \alpha_{l_a \kappa_a, l\kappa}^A \left(V_{l\kappa}^A + \sum_{B \neq A} \sum_{l_b \kappa_b} T_{l\kappa, l_b \kappa_b}^{AB} \Delta Q_{l_b \kappa_b}^B \right). \tag{7.25}$$

To solve these equations we can use a simple iterative approach by continued application of Equations (7.23) and (7.24) until the induced moments converge. This requires repeated evaluation of the potential which, depending on the system and the maximum value of l chosen, may be time consuming.

Alternatively, Equation (7.25) can be written as a matrix equation [37, 54, 55]

$$\mathbf{Y} \Delta \mathbf{Q} = \alpha \mathbf{V}_0, \tag{7.26}$$

where $\mathbf{Y}$ is a block matrix and the A, Bth block is given by

$$Y_{AB} = \begin{cases} \mathbf{I} & A = B \\ -\alpha^A \mathbf{T}^{AB} & A \neq B \end{cases}. \tag{7.27}$$

The vector $\Delta \mathbf{Q}$ contains the induced multipoles on the polarizable centers; α is a block diagonal matrix of the polarizabilities, and $\mathbf{V}_0$ is the vector describing the external field at each of the polarizable sites. This equation can be solved iteratively for the induced multipoles.

Inverting the polarizability, the equation becomes

$$\mathbf{Z} \Delta \mathbf{Q} = \mathbf{V}_0 \tag{7.28}$$

where

$$Z_{AB} = \begin{cases} (\alpha^A)^{-1} & A = B \\ -\mathbf{T}^{AB} & A \neq B \end{cases} \tag{7.29}$$

where $\mathbf{Z}$ is a symmetric matrix. This form of the equation is now amenable to more efficient matrix techniques, which may be important for large systems containing high-order distributed multipoles. As suggested by

Mazur [55], the method of conjugate gradients with Lanczos recursion may be used. Typically this does not require a large number of iterations to converge, and the most expensive part of each step is a simple matrix multiplication and hence does not require the repeated evaluation of the potential.

7.2.6 Damping

The multipole expansion is only valid when the charge distributions are not overlapping and when they are, divergence can occur. In our method any two-body divergence is removed, as monomers that are close together are evaluated using QM. The many-body induction energy, however, includes monomers that are close together and divergence can occur. Recently Slipchenko and Gordon [56] have reviewed damping functions within the framework of the effective fragment potential method for electrostatic, induction, and dispersion energies, and we direct the reader to this work and references therein for more details. In our work we have primarily considered damping within the induction calculation. Here the interaction tensors are replaced by damped tensors

$$T^{AB}_{l_a\kappa_a,l_b\kappa_b} = T^{AB}_{l_a\kappa_a,l_b\kappa_b} f_{l_a+l_b+1}(R), \tag{7.30}$$

where $f_n(R)$ is the damping function. Unfortunately it is not clear theoretically what the form of a damping function for the induction energy should be. A pragmatic approach is taken, using the function that performs best, as discussed by several authors [56–58]. Stone and coworkers [57–61] have used the Tang–Toennies damping function and we follow this. The Tang–Toennies damping function was derived [62] for the dispersion energy and assumes a short-range Born–Mayer repulsive potential. This derivation yields an incomplete gamma function and is given by

$$f_n(r) = 1 - \exp(-\delta r) \sum_{k=0}^{n} \frac{(\delta r)^k}{k!} \tag{7.31}$$

where n is the order of the function, r is the distance between the sites, and δ is a parameter that determines the length over which the damping is effective. In the original derivation the parameter δ was the exponential parameter in the Born–Mayer potential. Since the repulsive nature of the potential results from the exchange interaction, we can use the overlap model as mentioned in Section 7.2.4.5, and rewrite the Tang–Toennies function (in an analogous way to [56]) as

$$f_n(S_{AB}) = 1 - S_{AB} \sum_{k=0}^{n} \frac{(-\log(S_{AB}))^k}{k!} \tag{7.32}$$

where we have now removed any parameters. The overlap between densities can be calculated efficiently using density fitting. Another approach is to define δ in terms of the ionization potentials, I_M, of the two species $\delta = \sqrt{2I_A} + \sqrt{2I_B}$ [57–59]. And of course many approaches simply fit the parameters to reproduce some data, often using different parameters for different sites and orders n [63].

Since the same damping function can be used for dispersion energies for no extra computational cost we use it here as well, although it has little effect outside the QM region. The damped isotropic dispersion energy is given by

$$E_{\text{disp}} = -\sum_{n=6}^{10} f_n(R) \frac{C_n}{R^n}. \tag{7.33}$$

7.2.7 Periodic-boundary conditions

As mentioned in the introduction, periodic-boundary conditions (PBC) can be used to represent the bulk behavior of a system without having to include millions of particles in the simulation. The details of this are discussed in many textbooks such as [1]. When using PBC we need a method to include the effect of all the periodic images of the simulation cell, and for the electrostatic interaction this is usually done using the well-known Ewald summation. In the subsequent sections we will review how this is implemented for multipoles and also how it can be used in self-consistent induction calculations.

7.2.7.1 Multipolar Ewald summation

The original Ewald summation [64] was formulated for point charges but has been extended to dipoles and quadrupoles [65–68]. Most formulations have been in terms of Cartesian multipoles, but we prefer to follow that of Leslie using the spherical harmonic representation, which is applicable to arbitrary orders of multipoles [69]. In what follows we repeat several formulae from Leslie's work and rework them slightly to be more convenient for our implementation, but we refer the reader to the original paper for a more thorough discussion, as well as to [1] and [70] for a more general description of Ewald summation. The direct-space term as given by Leslie is

$$E_{\text{dl}} = \frac{1}{2} \sum_{AB} \sum_{l_a \kappa_a} \sum_{l_b \kappa_b} \binom{l_a + l_b}{l_a} Q_{l_a \kappa_a}^A Q_{l_b \kappa_b}^B \sum_{N}' \bar{S}_{l_a l_b (l_a + l_b)}^{\kappa_a \kappa_b} (\Omega_A, \Omega_B, \Omega_{AB_N})$$

$$\times \left\{ R_{AB_N}^{-l_a - l_b - 1} - \frac{R_{AB_N}^{l_a + l_b}}{(2l_a + 2l_b - 1)!!} \sqrt{\frac{2}{\pi}} (2\gamma)^{l_a + l_b + \frac{1}{2}} F_{l_a + l_b} \left(\gamma R_{AB_N}^2 \right) \right\}, \tag{7.34}$$

where the index N denotes a direct-space lattice vector, R_{AB_N} is the distance between molecules A and B, which may be in different unit cells separated by lattice vector N, and the prime on the summation indicates that interactions of particles with themselves should be neglected, i.e., when $R_{AB_N} = 0$. γ defines how broad the compensating potential is and $F_m(t)$ is the Boys function, given by

$$F_m(t) = \int_0^1 u^{2m} \exp(-tu^2) du. \tag{7.35}$$

Substituting the interaction tensor $\mathbf{T}$ into Equation (7.34) we get

$$E_{dl} = \frac{1}{2} \sum_{AB} \sum_{l_a \kappa_a} \sum_{l_b \kappa_b} Q^A_{l_a \kappa_a} Q^B_{l_b \kappa_b} \sum_N{}' T^{AB_N}_{l_a \kappa_a, l_b \kappa_b}$$

$$\times \left\{ 1 - \sqrt{\frac{2}{\pi}} \frac{R^{2l_a+2l_b+1}_{AB_N}}{(2l_a + 2l_b - 1)!!} (2\gamma)^{l_a+l_b+\frac{1}{2}} F_{l_a+l_b}\left(\gamma R^2_{AB_N}\right) \right\}. \tag{7.36}$$

The parameter γ is chosen such that the direct-space sum converges within a single unit cell; that is, only the $N = 0$ contribution is evaluated and this term can be computed in an analogous way to the electrostatic energy described in Section 7.2.4.6.

To determine the true unscreened electrostatic energy, the effect of this compensating potential must be removed and this is done through a reciprocal-space term. This is given by Leslie as

$$E_{rl} = \frac{4\pi}{V} \sum_{AB} \sum_{l_a \kappa_a} \sum_{l_b \kappa_b} \binom{l_a + l_b}{l_a} Q^A_{l_a \kappa_a} Q^B_{l_b \kappa_b} \frac{(-1)^{l_a+l_b}}{(2l_a + 2l_b - 1)!!}$$

$$\times \sum_n{}'' G(K_n) \cos^{(l_a+l_b)}(\mathbf{K}_n \cdot \mathbf{r}_{AB}) K^{l_a+l_b}_n \bar{S}^{\kappa_a \kappa_b}_{l_a l_b (l_a+l_b)} (\Omega_A, \Omega_B, \Omega_{K_n}) \tag{7.37}$$

where V is the volume of the unit cell, $\cos^{(l)}(z)$ the lth derivative of $\cos(z)$, $G(z) = \exp(-z^2/4\gamma)/z^2$, and $\mathbf{r}_{AB}$ the vector between sites A and B in the unit cell. The summation over n is over reciprocal-lattice vectors $\mathbf{K}_n$ with magnitude K_n and the double prime indicates that the reciprocal-lattice vector with magnitude $K_n = 0$ should be omitted. At this point Leslie performs a factorization over A and B, which we do differently because in our case Ω_A and Ω_B are 0. Rearranging and introducing the $\mathbf{T}$ tensors again

$$E_{rl} = \frac{4\pi}{V} \sum_{AB} \sum_{l_a \kappa_a} \sum_{l_b \kappa_b} Q^A_{l_a \kappa_a} Q^B_{l_b \kappa_b} \frac{(-1)^{l_a+l_b}}{(2l_a + 2l_b - 1)!!}$$

$$\times \sum_n{}'' G(K_n) \cos^{(l_a+l_b)}(\mathbf{K}_n \cdot \mathbf{r}_{AB}) K^{2l_a+2l_b+1}_n T^{K_n}_{l_a \kappa_a, l_b \kappa_b}. \tag{7.38}$$

We will use

$$\cos^{(l)}(A - B) = \sum_{\sigma=1}^{2} s_{\sigma,l} f_{\sigma,l}(A) g_{\sigma,l}(B),$$
(7.39)

$$s_{\sigma,l} = (-1)^{\lfloor (l+\sigma+1)/2 \rfloor},$$
(7.40)

$$f_{\sigma,l}(A) = \begin{cases} \cos(A) & \text{if } \sigma = 1, \\ \sin(A) & \text{if } \sigma = 2; \end{cases}$$
(7.41)

$$g_{\sigma,l}(B) = \begin{cases} \cos(B) & \text{if } \sigma + l \text{ is odd,} \\ \sin(B) & \text{if } \sigma + l \text{ is even;} \end{cases}$$
(7.42)

where $\lfloor x \rfloor$ is the floor function. Now substituting the Wigner rotation matrices for the interaction tensor gives

$$E_{\text{rl}} = \frac{4\pi}{V} \sum_{n}^{''} G(K_n) \sum_{l_a} \sum_{l_b} K_n^{l_a+l_b} \frac{(-1)^{l_a+l_b}}{(2l_a + 2l_b - 1)!!}$$

$$\times \sum_{\kappa} t_\kappa^{(l_a,l_b)} \sum_{\sigma=1}^{2} s_{\sigma,l_a+l_b} \left\{ \left(\sum_{A} f_{\sigma,(l_a+l_b)}(K_n \cdot r_A) \sum_{\kappa_a} Q^A_{l_a \kappa_a} D^{(l_a)}_{\kappa \kappa_a}(\Omega_{K_n}) \right) \right.$$

$$\times \left. \left(\sum_{B} g_{\sigma,(l_a+l_b)}(K_n \cdot r_B) \sum_{\kappa_b} Q^B_{l_b \kappa_b} D^{(l_b)}_{\kappa \kappa_b}(\Omega_{K_n}) \right) \right\}$$
(7.43)

where Ω_{K_n} is the orientation of the lattice vector K_n. The last two terms in parentheses are of the same form and thus they can be formed simultaneously, requiring only a single set of rotation matrices for all Q for each K_n.

The self-interaction term is exactly as in the Ewald summation for point charges

$$E_{\text{self}} = \sqrt{\frac{\gamma}{\pi}} \sum_{A} Q^A_{00}.$$
(7.44)

There is also the usual surface term, which depends on the total dipole moment of the cell and the dielectric constant, ϵ, of the system and is given by

$$E_{\text{surf}} = \frac{2\pi}{(2\epsilon + 1)V} \left| \sum_{A} Q^A_{00} r_A + \sum_{A} Q^A_1 \right|^2.$$
(7.45)

In addition, there are terms resulting from neglect of the reciprocal space term for which $K = 0$. In the traditional Ewald summation, this is divergent and is accounted for by the surface term. In multipolar Ewald summation, terms in which $l = 2$ do not diverge and have a finite value and must be included. This contribution is given by

$$E_{K=0} = \frac{2\pi}{3V} \left(\left| \sum_A \mathbf{Q}_1^A \right|^2 + 4 \left(\sum_A Q_{00}^A \right) \left(\sum_A Q_{20}^A \right) \right). \tag{7.46}$$

This term cannot be derived from Equation (7.43) as the rotation matrices are not defined for vectors of length zero, but we refer the reader to [69] for a more complete derivation of this term. If distributed multipoles have been used, the intramolecular interactions between sites on the same molecules, which have been included, must be removed.

7.2.7.2   Madelung potential for arbitrary multipoles

In order to evaluate the induction energy we must be able to evaluate the potential due to a periodic lattice of multipoles. The so-called *Madelung potential* for multipoles is given by

$$V_{\text{MP}}^{A, l_a \kappa_a}(\mathbf{r}_A) = V_{\text{dl}}^{A, l_a \kappa_a}(\mathbf{r}_A) + V_{\text{rl}}^{A, l_a \kappa_a}(\mathbf{r}_A) + \delta_{l_a 0} V_{\text{self}}^{A, 00}(\mathbf{r}_A) + \delta_{l_a 0} V_{\text{surf}}^{A, 00}(\mathbf{r}_A)$$

$$+ \delta_{l_a 1} V_{\text{surf}}^{A, 1\kappa_a}(\mathbf{r}_A) + \delta_{l_a 0} V_{K=0}^{A, 00}(\mathbf{r}_A) + \delta_{l_a 1} V_{K=0}^{A, 1\kappa_a}(\mathbf{r}_A)$$

$$+ \delta_{l_a 2} \delta_{\kappa_a 0} V_{K=0}^{A, 20}(\mathbf{r}_A) \tag{7.47}$$

where δ is the Kronecker delta symbol, and the various terms are given by

$$V_{\text{dl}}^{A, l_a \kappa_a}(\mathbf{r}_A) = \sum_B \sum_{l_b \kappa_b} Q_{l_b \kappa_b}^B \sum_N{}' T_{l_a \kappa_a, l_b \kappa_b}^{AB_N}$$

$$\times \left\{ 1 - \sqrt{\frac{2}{\pi}} \frac{R_{AB_N}^{2l_a + 2l_b + 1}}{(2l_a + 2l_b - 1)!!} (2\gamma)^{l_a + l_b + \frac{1}{2}} F_{l_a + l_b} \left(\gamma R_{AB_N}^2 \right) \right\} \tag{7.48}$$

$$V_{\text{rl}}^{A, l_a \kappa_a}(\mathbf{r}_A) = \frac{4\pi}{V} \sum_n{}'' G(K_n) \sum_\kappa D_{\kappa_a \kappa}^{(l_a)}(\Omega_K) \sum_{l_b} K_n^{l_a + l_b}$$

$$\times \frac{(-1)^{l_a + l_b}}{(2l_a + 2l_b - 1)!!} t_\kappa^{(l_a, l_b)} \sum_{\sigma=1}^2 S_{\sigma, l_a + l_b}$$

$$\times \left\{ f_{\sigma, (l_a + l_b)} (\mathbf{K}_n \cdot \mathbf{r}_A) \left(\sum_B g_{\sigma, (l_a + l_b)} (\mathbf{K}_n \cdot \mathbf{r}_B) \sum_{\kappa_b} Q_{l_b \kappa_b}^B D_{\kappa \kappa_b}^{(l_b)}(\Omega_K) \right) \right\} \tag{7.49}$$

$$V_{\text{self}}^{A, 00}(\mathbf{r}_A) = 2\sqrt{\frac{\gamma}{\pi}} Q_{00}^A \tag{7.50}$$

$$V_{\text{surf}}^{A,00}(\mathbf{r}_A) = \frac{2\pi}{(2\epsilon+1)V} \left(\mathbf{r}_A \cdot \left(\sum_B \mathbf{r}_B Q_{00}^B \right) + \mathbf{r}_A \cdot \left(\sum_B \mathbf{Q}_1^B \right) \right) \quad (7.51)$$

$$V_{\text{surf}}^{A,1\kappa_a}(\mathbf{r}_A) = \frac{2\pi}{(2\epsilon+1)V} \left(\sum_B \mathbf{Q}_1^B + \sum_B \mathbf{r}_B Q_{00}^B \right)_{\kappa_a} \quad (7.52)$$

$$V_{K=0}^{A,00}(\mathbf{r}_A) = \frac{8\pi}{3V} \sum_B Q_{20}^B \quad (7.53)$$

$$V_{K=0}^{A,1\kappa_a}(\mathbf{r}_A) = \frac{2\pi}{3V} \sum_B Q_{1\kappa_a}^B \quad (7.54)$$

$$V_{K=0}^{A,20}(\mathbf{r}_A) = \frac{8\pi}{3V} \sum_B Q_{00}^B. \quad (7.55)$$

7.2.7.3   Self-consistent polarization calculations in periodic boundary conditions

As discussed above, the induction energy must be calculated self-consistently and this must also be done in periodic boundary conditions. To do this we could evaluate the Madelung potential, $\mathbf{V}_{\text{MP}}$ due to the permanent multipoles, calculate the induced multipoles, calculate $\Delta\mathbf{V}_{\text{MP}}$ due to the induced multipoles, and so on. It may, however, become expensive to evaluate the Madelung potential repeatedly and it may also not be necessary. If the unit cell is large enough, it may suffice to calculate $\mathbf{V}_{\text{MP}}$ only once, and then at subsequent iterations calculate the potential due to the induced multipoles within the minimum image convention. This is likely to be adequate, as although the potential due to the permanent dipoles, for example, scales as R^{-3}, the potential due to the subsequent induced dipoles scales as R^{-6}. The effect of this approximation in a periodic box of 110 water molecules with side 14.89 Å, for example, is an error of 0.05 mE_h per water molecule. This is very dependent on the size of the unit cell, and if small cells are being used, the full potential should be employed throughout. If just the minimum image convention is used for the potential due to the induced multipoles, the matrix formulation described earlier can be used exactly as before, inserting $\mathbf{V}_{\text{MP}}$ for $\mathbf{V}_0$ into Equations (7.26) or (7.28).

7.2.8   Steps in an energy calculation

So far we have discussed all of the different pieces of the energy calculation and how they are evaluated. Here we summarize the steps actually performed in such a procedure:

1. Identify the monomers in the system and calculate the monomer energies E_1^{QM}, and the required multipoles and polarizabilities.

2. Determine dimer, trimer, etc., configurations that lie within the distance threshold and evaluate the QM energies to yield dimer, etc. interaction energies, $E_2^{QM,near}$, $E_3^{QM,near}$,....
3. Evaluate the exchange and dispersion energies for dimers not within the threshold, $E_2^{ex,far}$ and $E_2^{disp,far}$.
4. Evaluate the total electrostatic energy (Ewald summation if periodic), E_{tot}^{es}, and remove electrostatic energies of the dimers that have been calculated using QM, $E_2^{es,near}$. In nonperiodic systems, this energy difference is simply the electrostatic energy of the dimers outside the threshold, i.e., $E_{tot}^{es} - E_2^{es,near} = E_2^{es,far}$.
5. Calculate the many-body induction energy for the whole system, E_n^{ind}, and remove induction energies of the dimers, trimers, etc. that have been calculated using QM, $E_2^{ind,near}$, $E_3^{ind,near}$ ..., to prevent double counting.
6. The total energy can now be evaluated

$$E = E_1^{QM} + E_2^{QM,near} + E_3^{QM,near} + \cdots$$
$$+ E_2^{ex,far} + E_2^{disp,far} + \cdots$$
$$+ \left(E_{tot}^{es} - E_2^{es,near} \right)$$
$$+ \left(E_n^{ind} - E_2^{ind,near} - E_3^{ind,near} - \cdots \right). \tag{7.56}$$

7.2.9 Monte Carlo simulations

Liquids are not static collections of molecules and as such a single-point energy calculation of a particular configuration is not adequate. A more realistic description must sample the configurational space, and is accomplished using either Monte Carlo (MC) or molecular-dynamics simulations [1]. The latter require the evaluation of the forces on the molecules, a relatively straightforward extension of this work, but one that has not yet been implemented, so we choose to use MC methods. MC simulations involve making a random change to the configuration, evaluating the energy, and then accepting or rejecting this move with a probability related to a Boltzmann factor at the simulation temperature. Details of MC techniques can be found in [1] and [70] and will not be discussed here. In the Monte Carlo simulations used in our work, each random move involves only a single molecule and thus only properties and interactions associated with this molecule must be recalculated. The many-body induction calculation, however, must be performed at every step. Since only the closest interactions are calculated using expensive QM methods, and since only a single molecule is moved in each step, approximately the same number of interactions must be calculated at each step regardless of the size of the system under consideration. Thus the cost of a Monte Carlo step scales as $\mathcal{O}(1)$.

7.2.10   Parallel implementation

The most time-consuming step in the energy calculation will usually be the QM calculations on the monomers, dimers, trimers, etc. These tasks are completely independent of one another, so it is trivial to distribute them across multiple processors, and the energies and properties are simply collected at the end and processed. This type of parallelism can also be scaled up arbitrarily, so that as many processors as there are tasks can be used, or alternatively the tasks themselves (i.e., the QM calculations) can be parallelized to give a two-tier parallelism. The classical part of the calculation needs all of the monomer properties before it can begin, so this may influence the order in which the QM calculations are run. It may be possible to parallelize the many-body induction energy calculation, but this has not yet been carried out.

7.3   Examples

In all of the following examples, QM energy calculations are performed using DF-LMP2 [71] with the aug-cc-pVDZ basis set [72, 73]. The model potential comprises the electrostatic, induction, dispersion, and exchange energies. The electrostatic and induction energies have been calculated using MP2/aug-cc-pVDZ distributed multipoles and molecular polarizability tensors up to angular momentum $l = 3$. Dispersion energies are calculated using the isotropic expression given in Equation (7.33) and dispersion coefficients are taken from [74] and exchange energy is calculated using Equation (7.14) with $K = 0.3845\ E_{\mathrm{h}}$. Tang–Toennies damping is used for induction and dispersion energies and the overlap between monomers is calculated using the cc-pVTZ Coulomb density-fitting basis set [75] discarding any functions above d- and p-functions on oxygen and hydrogen, respectively. Distance thresholds are calculated between the centers of mass of the monomers. All calculations have been performed in a modified version of MOLPRO [76].

7.3.1   Two-body interactions

Figure 7.1 shows the QM and the model potential energy curves for the water dimer along the hydrogen-bond coordinate. As expected, the model potential does not reproduce the true curve exactly when the two molecules are close together, due to a very approximate treatment of the quantum-mechanical effects important at these separations. The quality of the model potential improves rapidly as the distance between the monomers increases and at distances just greater than 4 Å is indistinguishable from the QM curve. As such we have chosen 4.5 Å as our threshold distance, within which all dimer energies are treated with QM methods.

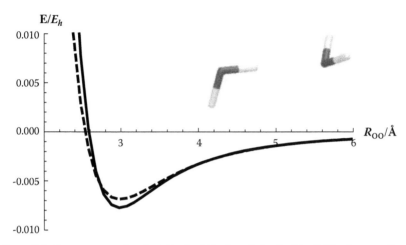

Figure 7.1 Quantum mechanical (dashed line) and classical (solid line) water dimer (configuration inset) potential energy curves for the oxygen–oxygen distance R_{OO}.

Figure 7.1 considers only a single dimer configuration, whereas in the liquid there are many other possibilities. To assess the performance of the two-body model potential for a selection of configurations, we have calculated the difference between the DF-LMP2/aug-cc-pVDZ energies and the model potential energies for all dimers in a $(H_2O)_{50}$ cluster (50 neighboring molecules taken from a much larger TIP4P water cluster). Figure 7.2 shows

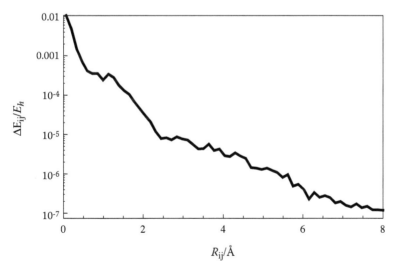

Figure 7.2 Average difference between the QM and model two-body interaction energies for all dimers in $(H_2O)_{50}$ with respect to their intermonomer distances.

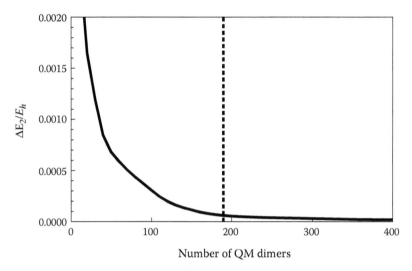

Figure 7.3 Total error in the two-body energy per water molecule in $(H_2O)_{50}$ with respect to the number of QM dimers included in the calculation. The dotted line indicates the number of QM dimers calculated within a threshold of 4.5 Å.

the average energy differences with respect to intermonomer separation (distances were sorted into bins of width 0.25 Å). By the threshold distance of 4.5 Å, we see that the average energy difference is of the order 10 μEh per water molecule. Figure 7.3 shows the error in the total two-body energy per water molecule as the number of QM dimers included in the calculation is increased. Approximately 190 of the 1225 possible dimers lie within the threshold 4.5 Å (marked by the dotted line) and it can be seen that this choice of threshold yields small errors in the two-body energy.

It should be noted that it is not obvious whether the DF-LMP2 or model energies are more accurate at distances where the model potential is valid. Given exact monomer properties, the model potential should give the exact interaction energy (where exchange can be completely neglected), but in this case we have neither exact interaction energies nor exact monomer properties. The differences are, however, small, which is the most important concern.

7.3.2 Three-body interactions

The model potential for three-body interactions consists solely of the induction energy. Three-body exchange and dispersion effects are neglected completely. In Figure 7.4 the average errors in the three-body energy per water molecule are shown for $(H_2O)_{50}$ with respect to the number of trimers included in the calculation. The criterion for categorizing trimers as being

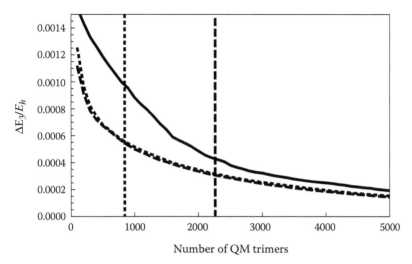

Figure 7.4 Average error in the total three-body energy per water molecule in $(H_2O)_{50}$ with respect to the number of QM trimers included in the calculation. The three lines correspond to three different criteria used to choose the trimers including: the minimum dimer distance (solid line), the two shortest dimer distances (dashed line), and all three dimer distances (dotted line). The vertical lines indicate where a threshold of 4.5 Å would come into effect. For the first criterion it is not shown but occurs when approximately 7800 trimers are included.

close is not as straightforward as that for dimers and there are several possibilities. Figure 7.4 shows the three possibilities for sorting the trimers and the error in the three-body energy per water molecule that results. The worst criterion is choosing trimers based on just the minimum distance between between any two of the dimers. This includes many trimers with the third molecule very distant from the others and approximately 7800 of the possible 19,600 trimers satisfy this condition in $(H_2O)_{50}$ with a threshold of 4.5 Å. Basing the selection on either two or three of the distances is more effective and the errors for both criteria scale similarly with increasing numbers of trimers, although there are far fewer trimers within the threshold if all three dimer distances are used.

7.3.3   Water clusters

Table 7.2 shows the errors in the computed energies of water clusters containing from 2 to 20 molecules. The first set of columns includes QM dimers and the second set includes dimers and trimers. For each we give the differences with respect to the total DF-LMP2/aug-cc-pVDZ energy: first, when neither a classical many-body contribution nor a threshold is used (ΔQM),

Table 7.2 Errors in Total Energies per Water
Molecule in mE_h for $(H_2O)_n$ $n = 2 - 20$

	Dimers			Trimers		
n	ΔQM	∞	4.5 Å	ΔQM	∞	4.5 Å
2	0.0	0.0	0.0	0.0	0.0	0.0
3	1.0	0.5	0.5	0.0	0.0	0.0
4	2.2	0.9	0.9	0.2	0.2	0.2
5	2.7	0.7	0.7	0.3	0.2	0.1
6	2.9	0.5	0.5	0.4	0.2	0.2
7	2.4	1.0	1.0	0.2	0.2	0.2
8	2.6	1.4	1.4	0.1	0.2	0.2
9	2.9	1.2	1.2	0.3	0.2	0.1
10	2.9	1.2	1.2	0.3	0.2	0.1
11	3.1	1.0	1.0	0.4	0.2	0.1
12	3.2	1.1	1.1	0.4	0.2	0.1
13	3.1	1.1	1.1	0.4	0.1	0.1
14	3.3	1.1	1.1	0.4	0.2	0.1
15	3.4	1.0	1.0	0.4	0.1	0.0
16	3.3	1.2	1.2	0.4	0.2	0.2
17	3.3	1.2	1.2	0.4	0.1	0.1
18	3.4	1.2	1.2	0.4	0.1	0.1
19	3.5	1.1	1.1	0.4	0.0	0.0
20	3.4	1.2	1.2	0.4	0.0	0.1

Note: ΔQM is the error in the QM energies, i.e., no classical many-body energy included. The columns
headed ∞ and 4.5 Å use no threshold and
a threshold of 4.5 Å for QM calculations. For
the three-body energies, all inter-monomer distances must be below this threshold.

and also when the many-body energy is included with (4.5 Å) and without
(∞) a QM threshold. The inclusion of the many-body energy in both the
dimer and trimer cases reduces the total error by approximately a factor
of 3. We also see that the use of a threshold of 4.5 Å does not significantly
affect the accuracy, that is, the model potential outside of this region is very
close to the DF-LMP2 potential.

7.3.4   Liquid water

Here we present some preliminary results for calculations on liquid water.
For a cell of 216 water molecules we have run constant pressure and con-
stant temperature Monte Carlo simulations using the standard Metropolis
Monte Carlo algorithm (see [70]). QM two-body interactions have been

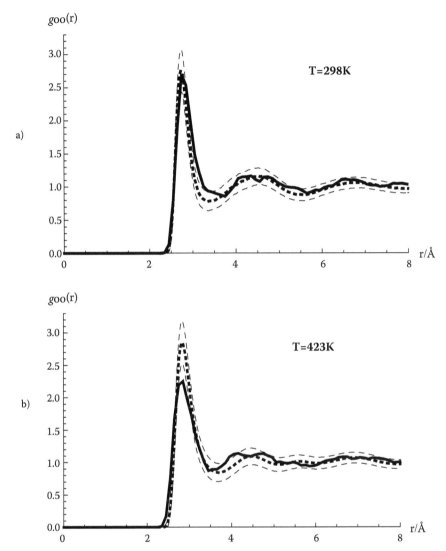

Figure 7.5 Oxygen–oxygen radial distribution functions, $g_{OO}(r)$, for water at different temperatures, T. The solid line is the computed RDF, the dotted line is the measured RDF, and the two lighter dashed lines indicate the experimental error.

calculated within a threshold of 4.5 Å and periodic boundary conditions for electrostatic and induction energies have been used. Approximately 100,000 steps have been performed in each simulation. Figure 7.5 shows the oxygen–oxygen radial distribution functions (RDF) from these simulations for different temperatures and pressures (corresponding to

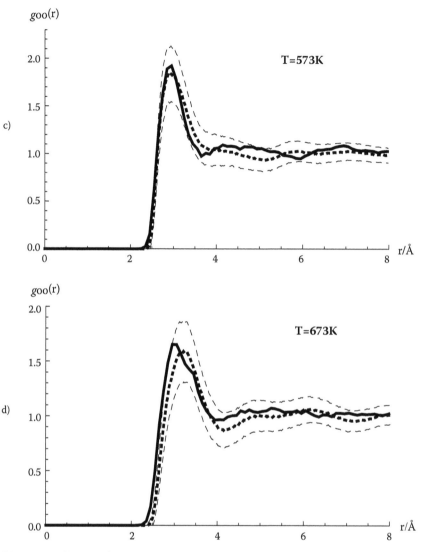

Figure 7.5 (*Continued.*)

298a, 423a, 573a, and 673a in [77]). The computed RDFs are compared to the experimentally measured RDFs [77] and their errors. The computed RDFs are still quite noisy as not enough Monte Carlo steps have been performed, but there is good agreement within the experimental errors. For the simulation at 298K we see particularly good reproduction of the structure of the RDF, and for the higher temperature structures we can see the features of the experimental RDFs emerging. These results are preliminary and it is clear that longer simulations must be run, but it demonstrates that it is indeed feasible to run ab initio calculations on liquids.

7.4 Conclusions

We have presented a conceptually simple method by which wavefunction-based methodologies can be applied to the condensed phases and have demonstrated this for liquid water. We have constructed our method by simplifying the problem into smaller, physically motivated pieces with the constraint that it should be systematically improvable. We have achieved this because we can always improve the quality of our simulations by increasing the method or basis set, by using higher-order QM interactions, by increasing the size of our threshold, or by using multipoles and polarizabilities of higher angular momenta. We have introduced a single, fitted parameter for estimation of the exchange energy, but this could be omitted, as in real simulations if the exchange energy is important, this interaction will be calculated explicitly using QM. Pragmatically our choice of damping function works well, but we are working on alternative solutions to the damping problem.

In this chapter we have shown results for water clusters and for liquid water. The ability to carry out such simulations opens the door to a whole host of possibilities, such as exploring the phase diagram of water, nucleation of water droplets, surface formation energies, and many other applications important throughout chemistry and we are currently starting work on such projects. Furthermore, this method is not limited to the liquid phase and is also applicable to molecular solids and we are currently using it to investigate different forms of ice.

In contrast to potential-based methods, which have been parameterized to specific systems, we can also apply this method to species other than water without any changes to our program. All of the data required, the multipoles, and polarizabilities, are generated from QM calculations on the fly so no parameterization is necessary. This allows, without significant additional effort, the simulation of processes involving multiple species, such as solvation, over wide ranges of temperature and pressure and offers the exciting possibility of new insights into a wide range of important, but still poorly understood, chemical phenomena.

References

[1] D. Frenkel and B. Smit, *Understanding molecular simulation: from algorithms to applications* (Academic Press, San Diego, 2002), 2nd edn.

[2] J. E. Jones, On the determination of molecular fields — II From the equation of state of a gas, *Proc. R. Soc. Lond. A* **106**, 463 (1924).

[3] R. Bukowski, K. Szalewicz, G. C. Groenenboom, and A. van der Avoird, Predictions of the properties of water from first principles, *Science* **315**, 1249 (2007).

[4] R. Bukowski, K. Szalewicz, G. C. Groenenboom, and A. van der Avoird, Polarizable interaction potential for water from coupled cluster calculations. I. Analysis of dimer potential energy surface, *J. Chem. Phys.* **128**, 094313 (2008).

[5] B. Jeziorski, R. Moszynski, and K. Szalewicz, Perturbation-theory approach to inter-molecular potential-energy surfaces of van-der-Waals complexes, *Chem. Rev.* **94**, 1887 (1994).

[6] B. Guillot, A reappraisal of what we have learnt during three decades of computer simulations on water, *J. Mol. Liq.* **101**, 219 (2002).

[7] W. L. Jorgensen and J. Tirado-Rives, Potential energy functions for atomic-level simulations of water and organic and biomolecular systems, *Proc. Natl. Acad. Sci.* **102**, 6665 (2005).

[8] R. G. Parr and Y. Weitao, *Density-functional theory of atoms and molecules* (Oxford University Press, New York, 1994).

[9] R. Dovesi, B. Civalleri, R. Orlando, C. Roetti, and V. R. Saunders, *Reviews in Computational Chemistry (Volume 21)* (Wiley-VCH, Weinheim, 2005).

[10] J. S. Tse, Ab initio molecular dynamics with density functional theory, *Annu. Rev. Phys. Chem.* **53**, 249 (2002).

[11] T. Helgaker, P. Jorgensen, and J. Olsen, *Molecular Electronic-Structure Theory* (Wiley, New York, 2000).

[12] M. Marsman, A. Grüneis, J. Paier, and G. Kresse, Second-order Moller–Plesset perturbation theory applied to extended systems. I. Within the projector-augmented-wave formalism using a plane wave basis set, *J. Chem. Phys.* **130**, 184103 (2009).

[13] C. Pisani, L. Maschio, S. Casassa, M. Halo, M. Schütz, and D. Usvyat, Periodic local MP2 method for the study of electronic correlation in crystals: Theory and preliminary applications, *J. Comput. Chem.* **29**, 2113 (2008).

[14] P. Y. Ayala, K. N. Kudin, and G. E. Scuseria, Atomic orbital Laplace-transformed second-order Møller–Plesset theory for periodic systems, *J. Chem. Phys.* **115**, 9698 (2001).

[15] W. L. Jorgensen, J. Chandrasekhar, J. D. Madura, R. W. Impey, and M. L. Klein, Comparison of simple potential functions for simulating liquid water, *J. Chem. Phys.* **79**, 926 (1983).

[16] H. J. C. Berendsen, J. R. Grigera, and T. Straatsma, The missing term in effective pair potentials, *J. Phys. Chem.* **91**, 6269 (1987).

[17] W. L. Jorgensen and C. Jenson, Temperature dependence of TIP3P, SPC, and TIP4P water from NPT Monte Carlo simulations: Seeking temperatures of maximum density, *J. Comput. Chem.* **19**, 1179 (1998).

[18] K. Laasonen, M. Sprik, M. Parrinello, and R. Car, "Ab initio" liquid water, *J. Chem. Phys.* **99**, 9080 (1993).

[19] P. Silvestrelli and M. Parrinello, Water Molecule Dipole in the Gas and in the Liquid Phase, *Phys. Rev. Lett.* **82**, 3308 (1999).

[20] E. Schwegler, G. Galli, and F. Gygi, Water under Pressure, *Phys. Rev. Lett.* **84**, 2429 (2000).

[21] I. Kuo, C. Mundy, M. McGrath, J. Siepmann, J. VandeVondele, M. Sprik, J. Hutter, B. Chen, M. Klein, F. Mohamed, M. Krack, and M. Parrinello, Liquid

water from first principles: Investigation of different sampling approaches, *J. Phys. Chem. B* **108**, 12990 (2004).

[22] J. C. Grossman, E. Schwegler, E. W. Draeger, F. Gygi, and G. Galli, Towards an assessment of the accuracy of density functional theory for first principles simulations of water, *J. Chem. Phys.* **120**, 300 (2004).

[23] J. VandeVondele, F. Mohamed, M. Krack, J. Hutter, M. Sprik, and M. Parrinello, The influence of temperature and density functional models in ab initio molecular dynamics simulation of liquid water, *J. Chem. Phys.* **122**, 014515 (2005).

[24] T. Todorova, A. Seitsonen, J. Hutter, I. Kuo, and C. Mundy, Molecular dynamics simulation of liquid water: Hybrid density functionals, *J. Phys. Chem. B* **110**, 3685 (2006).

[25] H.-S. Lee and M. E. Tuckerman, Structure of liquid water at ambient temperature from ab initio molecular dynamics performed in the complete basis set limit, *J. Chem. Phys.* **125**, 154507 (2006).

[26] H.-S. Lee and M. E. Tuckerman, Dynamical properties of liquid water from ab initio molecular dynamics performed in the complete basis set limit, *J. Chem. Phys.* **126**, 164501 (2007).

[27] T. D. Kuehne, M. Krack, and M. Parrinello, Static and dynamical properties of liquid water from first principles by a novel Car–Parrinello-like approach, *J. Chem. Theory Comput.* **5**, 235 (2009).

[28] B. Santra, A. Michaelides, and M. Scheffler, Coupled cluster benchmarks of water monomers and dimers extracted from density-functional theory liquid water: The importance of monomer deformations, *J. Chem. Phys.* **131**, 124509 (2009).

[29] E. E. Dahlke and D. G. Truhlar, Electrostatically embedded many-body correlation energy, with applications to the calculation of accurate second-order Møller–Plesset perturbation theory energies for large water clusters, *J. Chem. Theory Comput.* **3**, 1342 (2007).

[30] E. E. Dahlke and D. G. Truhlar, Electrostatically embedded many-body expansion for large systems, with applications to water clusters, *J. Chem. Theory Comput.* **3**, 46 (2007).

[31] H. R. Leverentz and D. G. Truhlar, Electrostatically embedded many-body approximation for systems of water, ammonia, and sulfuric acid and the dependence of its performance on embedding charges, *J. Chem. Theory Comput.* **5**, 1573 (2009).

[32] G. J. O. Beran, Approximating quantum many-body intermolecular interactions in molecular clusters using classical polarizable force fields, *J. Chem. Phys.* **130**, 164115 (2009).

[33] S. S. Xantheas, Ab initio studies of cyclic water clusters $(H_2O)_N$, $N = 1$-6. 2. Analysis of many-body interactions, *J. Chem. Phys.* **100**, 7523 (1994).

[34] M. P. Hodges, A. J. Stone, and S. S. Xantheas, Contribution of many-body terms to the energy for small water clusters: A comparison of ab initio calculations and accurate model potentials, *J. Phys. Chem. A* **101**, 9163 (1997).

[35] S. F. Boys and F. Bernardi, Calculation of small molecular interactions by differences of separate total energies—some procedures with reduced errors, *Mol. Phys.* **19**, 553 (1970).

[36] H.-J. Werner and K. Pflüger, *Annual reports in computational chemistry 2*, Volume 2 (Elsevier Science, Amsterdam, 2006).

[37] A. J. Stone, *The theory of intermolecular forces* (Oxford University Press, UK, 2002).

[38] A. J. Stone and M. Alderton, Distributed multipole analysis—methods and applications, *Mol. Phys.* **56**, 1047 (1985).

[39] A. J. Stone, Distributed multipole analysis, or how to describe a molecular charge-distribution, *Chem. Phys. Lett.* **83**, 233 (1981).

[40] A. J. Stone, Distributed multipole analysis: Stability for large basis sets, *J. Chem. Theory Comput.* **1**, 1128 (2005).

[41] R. J. Wheatley and T. C. Lillestolen, Local polarizabilities and dispersion energy coefficients, *Mol. Phys.* **106**, 1545 (2008).

[42] A. J. Stone, Distributed polarizabilities, *Mol. Phys.* **56**, 1065 (1985).

[43] A. J. Misquitta and A. J. Stone, Distributed polarizabilities obtained using a constrained density-fitting algorithm, *J. Chem. Phys.* **124**, 024111 (2006).

[44] W. Rijks and P. E. S. Wormer, Correlated van der Waals coefficients for dimers consisting of He, Ne, H_2, and N_2, *J. Chem. Phys.* **88**, 5704 (1988).

[45] W. Rijks and P. E. S. Wormer, Erratum: Correlated dispersion coefficients for dimers consisting of CO, HF, H_2O and NH_3, *J. Chem. Phys.* **92**, 5754 (1990).

[46] W. Rijks and P. E. S. Wormer, Correlated van der Waals coefficients .2. dimers consisting of CO, HF, H_2O, AND NH_3, *J. Chem. Phys.* **90**, 6507 (1989).

[47] A. J. Stone and C. Tong, Local and non-local dispersion models, *Chem. Phys.* **137**, 121 (1989).

[48] G. Williams and A. J. Stone, Distributed dispersion: A new approach, *J. Chem. Phys.* **119**, 4620 (2003).

[49] R. J. Wheatley and S. L. Price, An overlap model for estimating the anisotropy of repulsion, *Mol. Phys.* **69**, 507 (1990).

[50] B. I. Dunlap, J. W. D. Connolly, and J. R. Sabin, On some approximations in applications of $X\alpha$ theory, *J. Chem. Phys.* **71**, 3396 (1979).

[51] C. Hättig, Recurrence relations for the direct calculation of spherical multipole interaction tensors and Coulomb-type interaction energies, *Chem. Phys. Lett.* **260**, 341 (1996).

[52] J. Ivanic and K. Ruedenberg, Rotation matrices for real spherical harmonics. Direct determination by recursion, *J. Phys. Chem.* **100**, 6342 (1996).

[53] J. Ivanic and K. Ruedenberg, Rotation matrices for real spherical harmonics. Direct determination by recursion, *J. Phys. Chem. A* **102**, 9099 (1998).

[54] J. Applequist, J. Carl, and K. Fung, Atom dipole interaction model for molecular polarizability—application to polyatomic-molecules and determination of atom polarizabilities, *J. Am. Chem. Soc.* **94**, 2952 (1972).

[55] G. Mazur, An improved SCPF scheme for polarization energy calculations, *J. Comput. Chem.* **29**, 988 (2008).

[56] L. V. Slipchenko and M. S. Gordon, Damping functions in the effective fragment potential method, *Mol. Phys.* **107**, 1 (2009).

[57] A. J. Misquitta, A. J. Stone, and S. L. Price, Accurate induction energies for small organic molecules. 2. Development and testing of distributed polarizability models against SAPT(DFT) energies, *J. Chem. Theory Comput.* **4**, 19 (2008).

[58] A. J. Misquitta and A. J. Stone, Accurate induction energies for small organic molecules: 1. Theory, *J. Chem. Theory Comput.* **4**, 7 (2008).

[59] G. W. A. Welch, P. G. Karamertzanis, A. J. Misquitta, A. J. Stone, and S. L. Price, Is the induction energy important for modeling organic crystals?, *J. Chem. Theory Comput.* **4**, 522 (2008).

[60] C. Millot, J. Soetens, M. Costa, M. Hodges, and A. J. Stone, Revised anisotropic site potentials for the water dimer and calculated properties, *J. Phys. Chem. A* **102**, 754 (1998).

[61] C. Millot and A. J. Stone, Towards an accurate intermolecular potential for water, *Mol. Phys.* **77**, 439 (1992).

[62] K. T. Tang and J. P. Toennies, An improved simple model for the van der Waals potential based on universal damping functions for the dispersion coefficients, *J. Chem. Phys.* **80**, 3726 (1984).

[63] E. M. Mas, R. Bukowski, K. Szalewicz, G. C. Groenenboom, P. E. S. Wormer, and A. van der Avoird, Water pair potential of near spectroscopic accuracy. I. Analysis of potential surface and virial coefficients, *J. Chem. Phys.* **113**, 6687 (2000).

[64] P. P. Ewald, Die Berechnung optischer und elektrostatischer Gitterpotentiale, *Ann. Phys.* **369**, 253 (1921).

[65] H. Kornfeld, Die Berechnung elektrostatischer Potentiale und der Energie von Dipolund Quadrupolgittern, *Z. Phys.* **22**, 27 (1924).

[66] T. M. Nymand and P. Linse, Ewald summation and reaction field methods for potentials with atomic charges, dipoles, and polarizabilities, *J. Chem. Phys.* **112**, 6152 (2000).

[67] T. Laino and J. Hutter, Notes on Ewald summation of electrostatic multipole interactions up to quadrupolar level [J. Chem. Phys. 119, 7471 (2003)], *J. Chem. Phys.* **129**, 074102 (2008).

[68] A. Aguado and P. A. Madden, Ewald summation of electrostatic multipole interactions up to the quadrupolar level, *J. Chem. Phys.* **119**, 7471 (2003).

[69] M. Leslie, DL MULTI—A molecular dynamics program to use distributed multipole electrostatic models to simulate the dynamics of organic crystals, *Mol. Phys.* **106**, 1567 (2008).

[70] M. P. Allen and D. J. Tildesley, *Computer simulation of liquids* (Oxford University Press, UK, 1987).

[71] H.-J. Werner, F. R. Manby, and P. J. Knowles, Fast linear scaling second-order Møller–Plesset perturbation theory (MP2) using local and density fitting approximations, *J. Chem. Phys.* **118**, 8149 (2003).

[72] R. A. Kendall, T. H. Dunning Jr., and R. J. Harrison, Electron-affinities of the 1st-row atoms revisited—systematic basis-sets and wave-functions, *J. Chem. Phys.* **96**, 6796 (1992).

[73] T. H. Dunning Jr., Gaussian-basis sets for use in correlated molecular calculations .1. The atoms boron through neon and hydrogen, *J. Chem. Phys.* **90**, 1007 (1989).

[74] P. E. S. Wormer and H. Hettema, Many-body perturbation theory of frequency-dependent polarizabilities and van der Waals coefficients: Application to $H_2O - H_2O$ and Ar–NH_3, *J. Chem. Phys.* **97**, 5592 (1992).

[75] F. Weigend, A fully direct RI-HF algorithm: Implementation, optimised auxiliary basis sets, demonstration of accuracy and efficiency, *Phys. Chem. Chem. Phys.* **4**, 4285 (2002).

[76] H.-J. Werner, P. J. Knowles, R. Lindh, F. R. Manby, M. Schütz, et al., *Molpro, version 2008.3, a package of ab initio programs* (2008), see http://www.molpro.net.

[77] A. K. Soper, The radial distribution functions of water and ice from 220 to 673 K and at pressures up to 400 MPa, *Chem. Phys.* **258**, 121 (2000).

Index